T5-AEX-734

Problems in the Behavioural Sciences

GENERAL EDITOR: Jeffrey Gray

EDITORIAL BOARD: Michael Gelder, Richard Gregory, Robert Hinde, Jos Jaspars, Christopher Longuet-Higgins

Thirst

Thirst

Barbara J. Rolls and Edmund T. Rolls

University of Oxford
Department of Experimental Psychology
Oxford England

CAMBRIDGE UNIVERSITY PRESS

Cambridge
London New York New Rochelle
Melbourne Sydney

Published by the Press Syndicate of the University of Cambridge
The Pitt Building, Trumpington Street, Cambridge CB2 1RP
32 East 57th Street, New York, NY 10022, USA
296 Beaconsfield Parade, Middle Park, Melbourne 3206, Australia

© Cambridge University Press 1982

First published 1982

Printed in Great Britain by
Redwood Burn Limited
Trowbridge, Wiltshire

Library of Congress catalogue card number: 81—10206

British Library Cataloguing in Publication Data

Rolls, Barbara J.
Thirst.—(Problems in the behavioural sciences)
1. Thirst 2. Psychology, Physiological
I. Title II. Rolls, Edmund T.
III. Series
612′.391   QP139
ISBN 0 521 229189   hard covers
ISBN 0 521 29718 4   paperback

Contents

List of abbreviations

ADH, antidiuretic hormone (vasopressin)
AV3V, anteroventral region of the third ventricle
CSF, cerebrospinal fluid
ECF, extracellular fluid
ICF, intracellular fluid
OVLT, organum vasculosum of the lamina terminalis
SFO, subfornical organ
SIP, schedule-induced polydipsia

Foreword

The space in which science lives and grows has some curious features. It is natural in some ways to imagine a tree diagram: a thick trunk (philosophy?) divides first into a few major branches (physico-chemical, biological, social sciences), then into somewhat thinner branches (physics, for example, now separating from chemistry), and so on repeatedly until the twigs that display the latest, developing buds ramify in all their recondite glory. The trouble with this image is that it leaves no room for the fusions between long-separate branches which make up one of the most powerful driving forces in contemporary science. If biology and chemistry once split off from each other as branches that aim at different places in the sun, how can we put them together again in our tree diagram to symbolize that modern prodigy, biochemistry?

Another image, avoiding this problem, is that of the bicycle wheel: a central hub (philosophy?) with spokes radiating in all directions, the distance between spokes being proportional to the distance between the sciences they stand for. Now, to celebrate the birth of biochemistry, we merely have to slot in another spoke, equidistant between the parent disciplines. And, to preserve the necessary metaphor of growth that is integral to the tree diagram, we can make our spokes expand proportionally to the maturity of the subjects they represent. But it is a consequence of this image that spokes get further and further from each other as they grow: the quickest route between two disciplines is always through the philosophical hub; and, as subjects mature, they become more isolated, more idiosyncratic. To be sure, this feature of 'bicycle wheel' space is, to some extent, veridical: Bear witness the latest journal devoted entirely (I invent, but barely) to the eating habits of the *ob/ob* mouse after lesions to the ventromedial nucleus of the hypothalamus. But, from another point of view, it is the growing points of science that are the closest together, the circumference of the wheel that is shorter than the hub. To take an example close to our interests, it is those workers who push the frontiers of brain science forward most vigorously whose research may be transformed overnight by developments in physics, chemistry or linguistics; the linguists, whose subject may be revolutionized by advances in brain science.

The bicycle wheel is ill-designed for these paradoxes. Our series, *Problems in the Behavioural Sciences*, in contrast, is designed precisely to house them. Its intention is to provide a forum in which research workers on any frontier that relates to psychology *de facto*, even if not *de jure*, can communicate their discoveries, their questions and their problems to their peers in other disciplines, and to a new generation of students in their own. The central role played by psychology in this series guarantees that it will be multi-disciplinary; for psychology already draws on advances in fields as diverse as biophysics and biochemistry, logic and linguistics, social anthropology and sociology. Conversely, the increasing use of behavioural methods – often, alas, ill-understood and worse-applied – by scientists in other fields of biological or sociological enquiry means that they too need to keep a wary eye on how the battle rages at the frontiers of psychology. They will find in *Problems in the Behavioural Sciences* the latest war reports.

Thirst is by its very nature inter-disciplinary: indeed, it makes a mockery of the barriers erected between disciplines by history and accident. A correct fluid balance is vital for survival. Loss of water sets in train a whole host of physiological mechanisms that are intricately designed to preserve that balance. The study of these mechanisms is the domain of the physiologist, and especially the endocrinologist. But the only way in which, in the end, the animal deprived of water can put matters right is by behaving appropriately: it must find water – learn to find it, if necessary – and drink. That is the province of the psychologist. And the story does not stop there. A full understanding of the behaviour of the thirsty animal, and of the integration of that behaviour with the physiological response to water loss, requires an analysis of the brain mechanisms which control the behaviour that replenishes a water deficit. So now we are in the territory of the neuroscientist, that modern breed who, as often as not, is himself a hybrid between psychologist and physiologist.

Barbara and Edmund Rolls chart these complex waters with the ease and skill that can come only to those who are expert in all these various fields. Psychologists will find in the story they tell an easily accessible account of a topic that is central to the field of motivation – and this is itself central to the whole of psychology. Roughly speaking, motivational systems can be divided into those which obey principles of homeostasis (i.e., whose object is to maintain the equilibrium of the *milieu intérieur*) and those which do not conform in any obvious way to these principles. The emotion of fear is probably the best understood example of a motivational system belonging to the second of these categories; and fear will form the subject of a later monograph in our series. Thirst, in contrast, is the very type of a homeostatic motivational system. It has so far figured less prominently in the textbooks of physiological psychology than its more complex relative, hunger; but, as this book shows, it has yielded sufficiently far to physiological, behavioural and neural analysis as now to serve as an excellent general model for the way in which homeostatic motivational systems work. Thus a reader keen to know what is generally afoot in the field of motivation, no less than one with a more specific interest in the subject of thirst, will find here plenty of nourishment.

Preface

Drinking is essential for most terrestrial species and requires integrated behavioural responses to physiological stimuli and environmental demands. Because the stimuli which produce drinking can often be identified, drinking provides an opportunity to analyse behaviour in terms of its physiological and neurophysiological bases as well as at the psychological and ethological levels. We have written this book on thirst to bring together these different approaches. It is particularly important to have a clear and integrated understanding of thirst because disturbances in the balance of body water and electrolytes can occur in many clinical conditions.

This book has several aims. One is to provide a text useful to students of psychology, physiology, medicine, biology and zoology. Another aim is to consider not only the analysis of mechanisms that can produce drinking, but also the extent to which the different mechanisms underlie normal drinking, whether it occurs for example following water deprivation or spontaneously. For example, not only is drinking which can be produced by activation of the renin–angiotensin system analysed, but the role of activation of this system in normal drinking is also considered. A third aim is to make the discussion of thirst as relevant as possible to man, and therefore studies in non-human primates and man are included. A fourth aim is to introduce the investigation of subjective aspects of thirst, and studies in which these aspects have been investigated quantitatively are described. A fifth aim is to discuss clinical disorders of fluid intake in relation to our understanding of thirst mechanisms. With these aims, it is hoped that this book will be of interest not only to students, but to all those interested in the scientific study of thirst, and in the physiological and neural bases of behaviour. At this point historically in the study of thirst, we have tried to give a generally useful account of thirst and also, in the ways described above, to introduce themes which we believe will be important in the future development of the subject.

Thirst is defined and introduced in Chapter 1, and in Chapter 2 useful background information on fundamental aspects of the body fluids and their measurement is presented for those readers who have not previously covered this topic. After a historical review of the investigation of thirst in Chapter 3, Chapters 4, 5 and 6 describe the initiation, maintenance and termination of drinking. The neural and pharmacological controls of thirst are described in Chapters 7 and 8. We have included a separate chapter (Chapter 9) on clinical problems of thirst. Particular attention is paid in Chapter 10 to the factors that control spontaneous drinking. To encourage our readers to think about this subject further, we have pointed in Chapter 11 to some areas in which fruitful developments can be made.

To assist the reader, summaries of important points appear at the ends of most sections of this book.

We would like to thank all of our colleagues, including A. N. Epstein, J. T. Fitzsimons, J. G. Gibbs, J. G. G. Ledingham, H. Lind, R. W. Lind, D. J. McFarland, S. Maddison, G. J. Mogenson, D. J. Ramsay, and R. J. Wood, who have collaborated in many of the experiments described. Our research was supported by The Medical Research Council of Great Britain. D. J. Ramsay, R. J. Wood, and P. A. Phillips made many useful suggestions for improvement of the manuscript, and we thank them.

Oxford   B. J. Rolls
                                                                                          E. T. Rolls

1 Introduction

Thirst is a subjective sensation aroused by a lack of water. As a powerful and compelling sensation, it is perhaps only exceeded by the hunger for air and by pain. Associated with the sensation of thirst is the desire to drink water, and usually thirsty subjects report a dry feeling in the mouth and find that water tastes pleasant. Some of these different sensations were described vividly by the explorer Sven Hedin (1899) who, after a terrible journey across the western Taklamakan desert during which camels and men of his caravan succumbed for lack of water, finally struggled to safety:

I stood on the brink of a little pool filled with fresh, cool water – beautiful water.

It would be vain for me to try to describe the feelings which now over-powered me. They may be imagined – they cannot be described. Before drinking I counted my pulse: it was forty-nine. Then I took the tin box out of my pocket, filled it, and drank. How sweet that water tasted! Nobody can conceive it who has not been within an ace of dying of thirst. I lifted the tin to my lips, calmly, slowly, deliberately, and drank, drank, drank, time after time. How delicious! what exquisite pleasure! The noblest wine pressed out of the grape, the divinest nectar ever made, was never half so sweet. My hopes had not deceived me. The star of my fortunes shone brightly as ever it did.

I do not think I at all exaggerate, if I say that during the first ten minutes I drank between five and six pints. The tin box held not quite an ordinary tumblerful, and I emptied it quite a score of times. At that moment it never entered my head that, after such a long fast, it might be dangerous to drink in such quantity. But I experienced not the slightest ill effects from it. On the contrary I felt how that cold, clear delicious water infused new energy into me. Every blood-vessel and tissue of my body sucked up the life-giving fluid like a sponge. My pulse, which had been so feeble, now beat strong again. At the end of a few minutes it was already fifty-six. My blood, which had lately been so sluggish and so slow that it was scarce able to creep through the capillaries, now coursed easily through every blood-vessel. My hands, which had been dry, parched, and hard as wood, swelled out again. My skin, which had been like parchment, turned moist and elastic. And soon afterwards an active perspiration broke out upon my brow. In a word, I felt my whole body was imbibing fresh life and fresh strength. It was a solemn, an awe-inspiring moment.

Recently, we have been able to investigate quantitatively the sensation of thirst, and some of the other sensations associated with thirst (B. J. Rolls, Wood, Rolls *et al.*, 1980). Human subjects were asked to

How thirsty do you feel now ?
Not at all   Very thirsty

Fig. 1.1. An example of a visual analogue rating scale used by human subjects to provide a measure of thirst. The 10-cm line was marked at the position which corresponded to the degree of thirst felt.

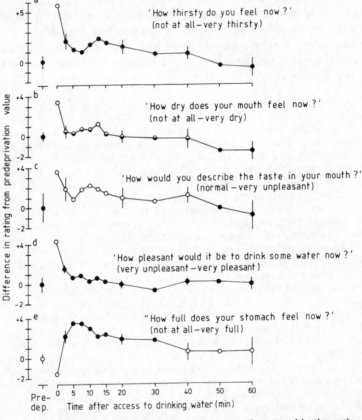

Fig. 1.2. The effect of 24 h water deprivation on human subjective ratings of thirst and other sensations is shown by the difference between the predeprivation rating (pre-dep.) and the rating at time 0, taken 24 h later just before access to water was given. The way in which the ratings changed after drinking started at time 0 is also shown. The significance of the changes relative to the value after 24 h water deprivation, at time 0, is indicated by closed circles ($P<0.01$), half-filled circles ($P<0.05$), or open circles (not significant). (After B. J. Rolls, Wood, Rolls *et al.*, 1980.)

respond to the question 'How thirsty do you feel now?' by marking the position corresponding to their sensation on a 10-cm line labelled at one end 'very thirsty' and at the other end 'not at all thirsty' (see figure 1.1). It is shown in figure 1.2a that, after 24 h deprivation (at time 0, just before access to water), the thirst rating moved more than 5 cm towards the 'very thirsty' end of the scale, compared to the normal, non-thirsty, rating. Water deprivation also usually produces a dry sensation in the mouth. Our subjects, consistent with this, demonstrated a significant shift in the rating of the dryness of the mouth (see figure 1.2b), and also reported a rather unpleasant, almost putrid, taste in the mouth, as shown in figure 1.2c. They also found the taste of water very pleasant after water deprivation, showing a significant shift towards pleasantness (figure 1.2d – see also Chapter 5). Associated with these sensations produced by water deprivation is a desire to drink water, and drinking after water is obtained is avid and remarkably rapid, as illustrated in figure 1.3. Interestingly, the sensation of thirst, and the dryness of the mouth and the pleasantness of the taste of water, diminish very rapidly once drinking is begun, with large and significant changes becoming evident within 2.5 min after starting to drink water (see figure 1.2).

As thirst is a subjective sensation aroused by a lack of water it can strictly only be studied directly in man, according to this definition. However, animals including man, when deprived of water, are in a state of drive in which they will search for and ingest water, and 'thirst' can be used in a different way to that described above as a name for this

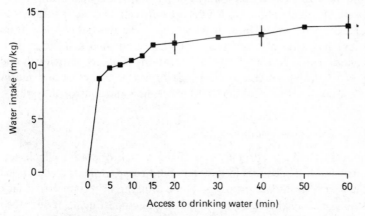

Fig. 1.3. The pattern of water intake over a 1-h period in which the 24-h water-deprived human subjects who provided the ratings shown in Figure 1.2 were allowed free access to drinking water. The drinking was remark-ably rapid. (Modified from B. J. Rolls, Wood, Rolls *et al.*, 1980.)

state of drive. Sometimes, the word thirst has also been applied to the deficits in body water produced by water deprivation, but this use is best avoided, as it is clear and accurate to refer in this case simply to 'water deficit'. The way in which body water is divided into different compartments and the influence which water deprivation has on these compartments, is described in Chapter 2.

The study of thirst and drinking

Thirst, including the control of drinking in animals and man, is the subject of this book. It is a useful and interesting subject to study, not only because it is essential to the survival of most terrestrial vertebrates, and is relevant to clinical conditions of water imbalance in man, as described in Chapter 9, but also because it provides a useful model system with which to analyse how a relatively complex type of behaviour is controlled. The advantages of using drinking as a model system are many. First, it is relatively easy to measure. Even with a behaviour as apparently similar as feeding, measurement is much more complicated in that allowance must be made for the ingestion of a wide range of nutrients, each of which has different effects on the system. Second, with drinking, the initiating stimuli are likely to be related to the effects which lack of water has on the system, and these can be measured and identified relatively straightforwardly. In the case of some other types of behaviour, it may be more difficult to identify and thus to analyse the signals that initiate the behaviour. For example, with feeding it is difficult to know which of the many chemical changes that are part of the metabolism of different nutrients may act as signals to control the behaviour. Third, drinking allows the analysis of body–brain relations in a case of a relatively complex, motivated behaviour. This behaviour can be related to the sensations that accompany it and analysed in terms of the whole range of sensory, control and motor processes which all contribute to the expression and control of drinking.

The causes of drinking

Thirst and drinking normally arise from a lack of water, which acts through resulting changes in the body fluid compartments as described in Chapter 4 to initiate drinking. Thus, drinking in response to a lack of water, or to an alteration in the body fluid compartments, is described as homeostatic in that it reduces the disturbances in the body fluid compartments. Homeostasis is an important concept used by W. B. Cannon (1947) to describe 'the various physiological arrangements which serve to restore the normal state once it has been disturbed'. Drinking in response to a lack of water is an example of a behavioural response which serves to maintain homeostasis. The controls of homeostatic (or 'primary') drinking, which occurs in response to body fluid imbalances, are described in Chapters 4, 5 and 6 in terms of the factors that initiate, maintain, and terminate the drinking. It will be seen that it is possible to specify quite precisely for drinking behaviour which signals lead to the drive for water, which signals reward or reinforce drinking, and which signals stop drinking. It is very interesting that different signals are involved in controlling these different aspects of behaviour. For example, drinking may be initiated in response to a body fluid imbalance, but may be rewarded in the first instance by the taste of water. As a basis for understanding the homeostatic controls of drinking, the fundamental aspects of body fluid distribution and balance are described in Chapter 2.

If drinking is not caused by a change in the body fluid compartments or occurs heedless of the state of body fluids, then it is described as non-homeostatic, in that it does not reduce physiological imbalances. This drinking has also been called non-regulatory drinking, or 'secondary' drinking. Examples of non-homeostatic drinking, that is, drinking which is inappropriate given the state of the body fluids, are as follows: (1) The under-drinking which occurs when fluids are made less palatable, for example by the addition of the bitter-tasting substance quinine. (2) Prandial drinking, which is drinking of small draughts taken alternately with small morsels of food in rapid succession in a meal (see Kissileff, 1973). This drinking is at least partly lubricative, and does not just serve to reduce the dehydrating consequences of the meal, in that it occurs primarily when salivation is experimentally reduced, either pharmacologically or by removal of the salivary glands. (3) A dry mouth alone, induced for example by the above methods, or by speaking, excessive smoking, panting or irritation by spicy foods (see Fitzsimons, 1979) or by the removal of the salivary glands (Kissileff, 1973), is sufficient to lead to increased drinking or to an altered pattern of drinking. (4) Schedule-induced polydipsia, in which a rat (reduced to 80% body weight by food restriction), when given 45 mg pellets of food on a spaced-reward schedule with water freely available, drinks excessively between the deliveries of food (Falk, 1961). The effort required to obtain water is another non-homeostatic control of drinking and can affect the drinking pattern as well as the amount ingested (see McFarland, 1971). For example, if the path between the places where food and water is obtained is made longer, then the duration of each bout of drinking and eating becomes longer. Non-homeostatic or secondary drinking is considered further in Chapter 10, together with the question of whether normal, everyday, drinking is in response to body fluid deficits and is thus homeostatic, or whether it anticipates these deficits so that they do not occur, or is in such great excess that it is far greater than that required for homeostasis.

Measures of drinking

A simple way to measure drinking which is widely used in the laboratory is to use an inverted graduated cylinder or tube with a drinking spout at the bottom (figure 1.4). As it allows total intake to be measured, and to be compared perhaps to a fluid-deficit signal, this method is particularly appropriate for determining how accurately drinking is terminated in relation to need. This is thus not so much a measure of how much thirst is being experienced or of the drive to drink, but more a measure of how much water is required to produce satiety, or at least to terminate drinking.

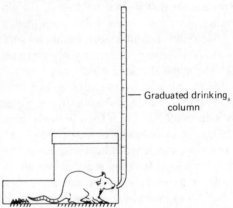

Graduated drinking, column

Fig. 1.4. An inverted graduated tube with a drinking spout is a simple way to measure drinking.

If a measure of thirst or the drive for water is required, it is better to measure how hard an animal will work to obtain water, rather than how much water must be consumed before satiety mechanisms terminate drinking. For example, in a progressive ratio test of the animal's motivation, one progressively increases the number of times the animal must press a lever in order to obtain one small delivery of water until the animal stops working. In one study of this type, a progressive ratio of 10 was used, so that rats had to press once for the first water reward, 11 times for the second, 21 times for the third, etc., and it was shown that the hormone angiotensin produced a motivation for water comparable to that produced by moderate water deprivation, in that the rats pressed up to approximately 51 times for one water reward in each of these conditions before they stopped work (B. J. Rolls, Jones & Fallows, 1972). Another example of a schedule in which motivation can be assessed without the complication of satiety produced by ingestion is a variable-interval schedule, in which rewards

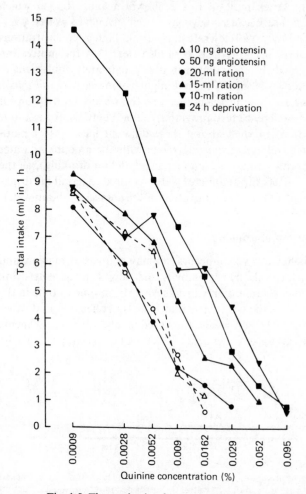

Fig. 1.5. The motivation for water can be estimated by the concentration of bitter-tasting quinine tolerated in the drinking water. In this case rats tolerated the highest concentrations of quinine after 24 h water deprivation, or if they were allowed only a 10-ml ration of water overnight (see text for further details). (From B. J. Rolls, Jones & Fallows, 1972.)

are delivered for the first press made after a variable interval of mean value, e.g. 1 min. On this type of schedule animals work steadily, and at a rate which gives a measure of their motivation, in that the rate is increased by deprivation. Another measure of motivation for water is to assess to what extent animals will tolerate aversive consequences in order to obtain water, using for example electric shock (Warden, 1931), or bitter-tasting quinine added to the water (e.g. B. J. Rolls, Jones & Fallows, 1972). An example of this is shown in figure 1.5, in which the thirstier rats are because they have been given less water overnight, the more they will tolerate and drink higher concentrations of quinine in a 1-h drinking test. It is also clear that the motivation for water of animals given angiotensin is approximately equivalent to that of rats allowed 15–20 ml of water overnight (compared to the *ad libitum* overnight intake of 25–45 ml). Another way to measure motivation for water is to use a preference test, providing a choice between for example water and food. As implied above, these different measures of motivation for water will not always correlate with the amount of water actually drunk with free access (known as *ad libitum* drinking), as the amount drunk reflects the amount of water required to terminate drinking, rather than just the initial motivation of the animal to obtain water.

The pattern of drinking

Drinking is a behaviour which occurs in many animals at regular, and quite frequent, intervals. Farm animals and dogs kept as pets drink several times a day. Man not only drinks water from time to time in the day, especially when the weather is hot, but also coffee, tea and other beverages, particularly with meals. Rats used in laboratory experiments and allowed free access to water drink at frequent intervals, particularly,

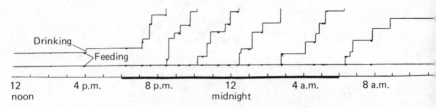

Fig. 1.6. Drinking and feeding patterns in a rat. Each time the rat licked the drinking tube to obtain water, the upper pen moved upward a fraction of a millimetre. (It reset at the top of the paper.) Each time the rat pressed a bar to obtain a 45-mg pellet of food, the upper and lower pens deflected downwards briefly, then returned. Note that most of the drinking was at night, was in long draughts, and was associated with food, occurring just before or just after eating. (From Epstein, 1967.)

as shown in figure 1.6, at night and in association with feeding. They usually lap at a relatively constant rate of 7 laps/s in short bouts of 30 s to 5 min, and intersperse these bouts with other behaviours such as feeding, grooming or resting. Although one or several bouts of drinking usually follow, or sometimes precede, a bout of feeding and the drinking is thus associated with feeding, it should be noted that this is different from the prandial drinking noted above in which the feeding and drinking are continually intermingled within a bout in desalivate animals (Kissileff, 1973). Prandial drinking appears to occur partly for lubrication, in that it is attenuated if water is infused into the mouth, but not if water is infused into the stomach. However, it is more difficult to know whether drinking that is normally associated with feeding occurs for lubrication or because of actual or anticipated dehydration produced by the food, or for other reasons. This important question, of whether normal drinking is often associated with feeding because dehydration arises or is anticipated as a result of the ingested food, is dealt with in Chapter 10.

After water deprivation, animals drink to satiety with different time-courses. Dogs typically drink very rapidly, gulping down enough in 2–3 min to replace eventually their body needs after 24 h water deprivation. Man and monkeys also drink quite rapidly after 24 h water deprivation, drinking much of what they need in 2.5 min, and most of it in 5 min (see figure 6.2). Rats are slower drinkers, continuing to drink significant quantities in the second half of a 1-h test session following 24 h water deprivation. These different patterns of drinking have important implications for the mechanisms by which drinking is terminated. In relatively slow drinkers, such as the rat, there may be sufficient time for water to be absorbed before drinking is terminated, so that simple replacement of the body fluid deficit may be sufficient to account for the termination of drinking. In species which drink relatively rapidly, there may be insufficient time for water to be absorbed by the gut to allow the deficit signals to be neutralized, so that specialized, pre-absorptive satiety systems are required. It is important to separate and analyse these different possibilities, as shown in Chapter 6.

Another feature in the pattern of drinking is a circadian rhythm. In the rat for example most of the drinking occurs at night (figure 1.6), and this rhythm persists even when food is not available (Morrison, 1968). It is not yet known whether this reflects a circadian rhythm in the drinking mechanism itself or whether the increased drinking at night could be due to a homeostatic need for water, resulting from the water lost in the saliva which is used for grooming as well as for behavioural thermoregulation in the rat (Kissileff, 1973).

2   Fundamentals of fluid intake and output

Life originated in the sea. In this environment the simplest strategy was for animals to evolve with an internal composition similar in concentration to that of the sea water surrounding them. However, as animals colonized dry land they retained a sea-like internal composition and therefore had to meet the problems of conserving and obtaining water. The way in which some terrestrial mammals, including man, have coped with these problems forms the basis of this book. Before attempting to understand the complexities of behavioural regulation it is necessary to explain the fundamental features of fluid composition and balance. The ways of measuring and manipulating the body fluids will also be introduced as these procedures form the backbone of our experimental understanding of the controls of thirst.

Water content of the body

The functions of all the cells of the body depend on water. Oxygen and nutrients are taken up from water and wastes are discharged into it. Water is the largest single component of the body and its volume must be maintained within narrow limits. In the nineteenth century Claude Bernard attempted to define the proportion of water in the body by comparing the weights of the desiccated bodies of Egyptian mummies with the weights of living persons of the same general size and shape. His estimate was that the body is 90% water, which we now know from the more direct method of drying the tissues of corpses is too high. The proportion of body water in man can vary widely in individuals and ranges from about 45% to 70%. Within an individual the proportion of body water tends to decrease with age. This is partly because the proportion of water decreases as the amount of body fat increases. The proportion of water to the lean body mass (the body without fat) is essentially constant at 70%. The proportion of body water and the distribution of water in the various organs tends to be similar in all terrestrial animals.

Total body water can now be measured by injecting a non-toxic substance (e.g. antipyrine, deuterium oxide or tritiated water) which distributes evenly throughout the body water. The volume of fluid in which the substance is distributed is then calculated by dividing the amount injected (minus any lost through excretion or metabolism) by the concentration of the substance in a plasma sample. The substance injected therefore must be easy to measure, either chemically or by counting a radioactive label (for further details, see Ganong, 1979).

Organization of the body fluids

The water of the body is organized into two main compartments, that inside the cells (intracellular or cellular), and that outside the cells

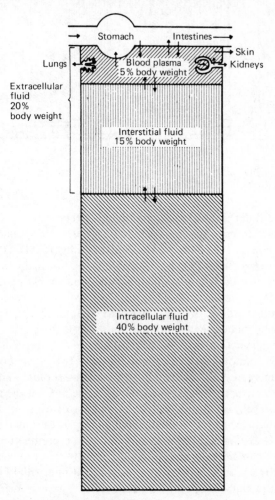

Fig. 2.1. Body fluid compartments. Arrows represent fluid movement. (From Gamble, 1954.)

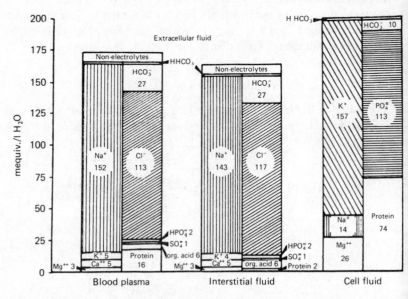

Fig. 2.2. Electrolyte composition of human body fluids. (From Leaf & Newburgh, 1955.)

(extracellular). The extracellular fluid is either in the vasculature (blood plasma) or between the cells (interstitial). The greatest proportion (two-thirds) of the fluid is intracellular (see figure 2.1). Only one quarter of the extracellular fluid is in the plasma, but the maintenance of plasma volume is vital for survival. Loss of blood or plasma water leads to circulatory collapse and death. We will see later that various mechanisms have evolved to ensure that both plasma volume and cellular volume are maintained.

The distinction between intracellular and extracellular fluid is not merely anatomical. The fluid in the two compartments differs fundamentally in composition. Most of the sodium and chloride ions are in the extracellular compartment, while most of the potassium ions are intracellular. Some protein anions are distributed throughout the body fluids, but the concentration is low in the interstitial fluid (see figure 2.2). These differences in composition are due to the properties of the cell membrane and the capillary walls. The capillary walls, which act as the barrier between the plasma and interstitial fluid, in general permit movement of all constituents of the plasma except protein. The cell membrane, which acts as the barrier between the cellular and extracellular fluid compartments, is not only impermeable to proteins, but normally also prevents the movement of potassium ions out of the cells. The cell membrane is not simply an inert barrier, but actively maintains the differences in concentration of sodium and potassium through metabolic processes. Various factors affect the movement of substances across the membrane (see Ganong, 1979), but the most important principle for the understanding of thirst is that of osmosis. When a membrane is impermeable to a substance in solution (solute), such as sodium, but permeable to the solvent (e.g. water), the solvent will tend to distribute so that the concentration of solute is the same on both sides of the membrane. This movement of solvent across a semi-permeable membrane into the more concentrated solution is called osmosis. The tendency for the solvent to move across the membrane can be prevented by applying pressure to the more concentrated

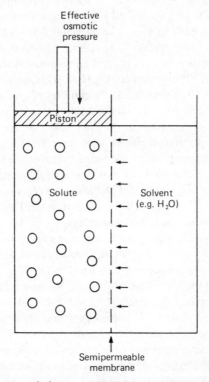

Fig. 2.3. When two solutions are separated by a semipermeable membrane, the solvent (e.g. water) tends to move from the more dilute solution to the more concentrated solution (i.e. the one with the most solute). If the membrane is rigid, the solvent movement can be prevented by exerting pressure on a piston, equal to the difference in osmotic pressure between the two solutions. The pressure required is the effective osmotic pressure.

solution. The amount of pressure needed to prevent movement is called the effective osmotic pressure of the solution (see figure 2.3).

Osmotic pressure may be measured in an osmometer which records the reduction in freezing point of a given solution. This depression of freezing point depends on the total number of osmotically active particles in a given volume of solution. So if, for example, sodium chloride completely dissociated into sodium ions and chloride ions, each mole (i.e. the molecular weight of a substance in grammes) in solution would produce 2 osmoles of osmotically active particles. If this amount of sodium chloride were dissolved in 1 kg of water, the osmolal concentration would be 2 osmoles/kg (of water). The body fluids, however, do not behave in this straightforward way. Many of the solutes do dissociate almost completely, but some of the ions interact, thereby reducing the number of particles in solution and the osmotic effect. Thus, if we counted all the particles which should be in solution in plasma the number would be over 300, but because of interactions in the solution the number is smaller; measurement of freezing point depression indicates that plasma concentration in a normal person is approximately 290 mosmoles/Kg H_2O. This is usually referred to as the osmolality of the solution, i.e. the number of osmoles per kg of solvent, and expresses the total activity of all the particles in solution.*

When the concentrations of other solutions are compared to that of plasma, the term tonicity is used to refer to the effective osmotic pressure developed across the semipermeable membrane. Plasma is the easiest body fluid to withdraw for analysis, and under steady-state conditions its osmolality will be the same as that of the other body fluids such as the cell contents. Tonicity is thus used to refer to the pressure developed across the cell membrane by a solution. If a solution has the same effective osmotic pressure as the cell contents, no diffusion of solvent (water) across the cell membrane occurs and the solution is said to be isotonic (a solution of sodium chloride of approximately 0.9% by weight or 0.15 M is isotonic). If a solution has a greater effective osmotic pressure than the cell contents and water diffuses out of the cell, causing cell shrinkage, it is said to be hypertonic. If a solution has a lower effective osmotic pressure than the cell contents and water diffuses into the cell, causing expansion of the cell, it is hypotonic. It should be noted that substances which cross the cell membrane freely, such as urea, do not contribute to the effective osmotic pressure developed across the cell membrane and do not affect the tonicity of the solution in which they are dissolved, although of course they do make a

* The term osmolarity will sometimes be encountered, which refers to the number of osmoles per litre of solution.

contribution (which is usually small) to the absolute osmotic pressure as measured in an osmometer.

Manipulating the fluid compartments

As will be explained in detail in later chapters, thirst depends on the volume and composition of the fluid compartments. Some of the standard ways of predictably altering the fluid compartments are shown in figure 2.4. Where appropriate more detail will be given in later chapters. In general it is easier to deplete or expand selectively the extracellular than the cellular compartment because isotonic fluid can be added to or taken away from the body without affecting the cells (see figure 2.4 and Chapter 4). In contrast, changes in body water which affect cellular volume and composition also influence the extracellular compartment, since water distributes freely across the cell membrane (see for example a, b and e in figure 2.4).

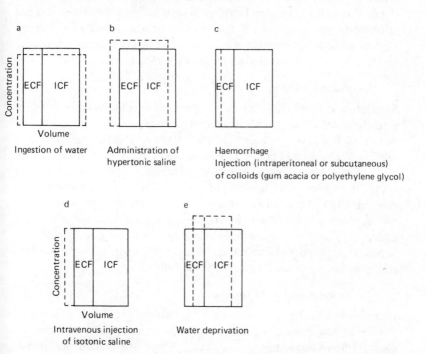

Fig. 2.4. Some ways in which the extracellular (ECF) and intracellular (ICF) fluid compartments can be manipulated. The solid lines represent the normal isotonic state. The dotted lines indicate the direction of changes in volume and composition (but not necessarily the magnitude) when equilibrium has been reached after the manipulation.

Measuring changes in the fluid compartments

Osmotic pressure or osmolality

As has been mentioned already, this measure indicates the overall concentration of the body fluids. This is one of the measures given most frequently in studies of thirst since, in general, an elevation of plasma osmolality will lead to a loss of cellular water (cellular dehydration), which is a powerful thirst stimulus. It is not, however, always the most useful measure. The freezing point is affected by the concentration of a variety of solutes, for example plasma protein and urea. As we shall see later it is the effective osmotic pressure or the ability of solutes in the plasma to withdraw water from the cells which is critical for the stimulation of drinking. Thus, a change in the depression of the freezing point of plasma is usually a good indication of effective osmotic pressure but if, for example, the concentration of urea is elevated, freezing point depression will indicate an increase in osmotic pressure. However, the effective osmotic pressure will not be changed since urea crosses the cell membrane and distributes evenly across it, so that it does not contribute to the effective osmotic pressure across the membrane.

Plasma sodium

Sodium is the principal cation of the plasma and together with its attendant anions accounts for approximately 90 to 95% of the osmotically active particles in the plasma. Plasma sodium and osmolality are usually highly correlated. Sodium can be measured easily and reliably by means of a flame photometer which works on the principle that sodium changes the colour of a flame according to its concentration. This colour change is measured and converted into milliequivalents of sodium in solution. Changes in plasma sodium concentration are the best indicators of the degree of cellular dehydration, as they reflect relatively accurately changes in the osmotic stimulus that withdraws water from the cell.

Extracellular fluid volume

Extracellular fluid (ECF) volume is measured by dilution techniques similar to those described for total body water. ECF is difficult to measure because there is no substance which will remain just in the extracellular compartment and distribute rapidly and uniformly throughout the ECF. Because of this problem the volume of distribution of a selected substance is measured. This measurement, while it does not necessarily give the absolute volume of ECF, is a useful indicator of changes in the ECF. Two general types of substances are used for determining the volume of distribution of the ECF:

(1) saccharides: e.g. inulin, sucrose, mannitol
(2) ions: e.g. thiocyanate, chloride, sodium, thiosulphate, sulphate, bromide

Plasma volume

Plasma volume is also measured by determining the dilution of a substance, such as a dye (Evans blue) which binds to plasma protein, or serum albumin labelled with radioactive iodine.

Haematocrit (or packed cell volume) is the measurement most often given in publications on drinking to indicate plasma volume changes. Haematocrit indicates the percentage of the total blood volume occupied by the red cells. It can be determined easily by centrifuging a small sample of blood (taken under free-flow conditions from a large vein) in a capillary tube and measuring the height of the layer of red cells which have been spun to the bottom. Although haematocrit does not indicate absolute plasma volume it can give an indication of changes in volume. For example, a decrease in plasma volume would be indicated by an increase in haematocrit since the number of red cells remains constant except during blood loss. Haematocrit values are sometimes quite variable, particularly in large animals, and do not always provide an accurate assessment of plasma changes.

Measurements of plasma protein give an indication of relative changes in plasma volume, since protein is usually retained when plasma water is lost, and generally are more reliable than the haematocrit. Plasma protein can be estimated by determining the specific gravity of plasma, by measuring its refractive index, or by chemical analysis.

Intracellular fluid volume

Intracellular fluid volume (ICF) cannot be measured directly, but can be calculated by subtracting the ECF volume from total body water. The best indication of changes in cellular hydration is given by changes in plasma sodium. For example, if plasma sodium concentration is high, water is withdrawn from the cells by osmosis and intracellular volume is reduced.

Gain and loss of body water

The fluid in the body has a particular volume and composition which remain relatively constant, despite the continuous and rapid turnover

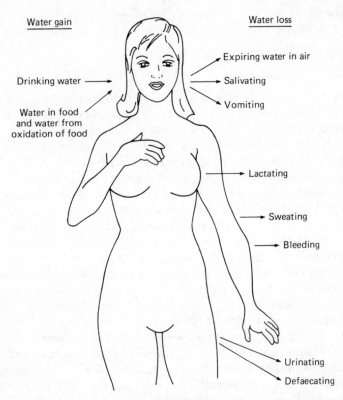

Fig. 2.5. Some ways in which water can be exchanged with the environment.

of water in the body. Some of the ways in which water can be exchanged with the environment are shown in figure 2.5. In man the turnover of water is usually about 2 or 3 l/day. Most body water comes from the ingestion of fluid, and it is the regulation of this drinking which forms the basis of this book. Fluid can be lost from the body in many ways and the importance of these different routes may vary considerably depending on the type and condition of the animal and on its environment. In general, the most important source of fluid loss is the urine. The kidney has evolved to filter the blood and to keep its volume and composition relatively constant by conserving the water and solutes needed and excreting any excess. The kidney, through its ability to conserve water, is the organ which has given terrestrial mammals the freedom to move about in their environment away from the waterhole.

The kidney

It is pointless to try and understand the controls of fluid intake without some basic understanding of the way in which fluid balance is maintained by the kidney since normally there is a close relation between intake and output.

It is the job of the kidney to form urine from the blood and thereby maintain the volume and composition of the blood. To this end the kidney receives a high blood flow, equivalent to some 25% of the output of the heart. Not only does the kidney form urine, but it also

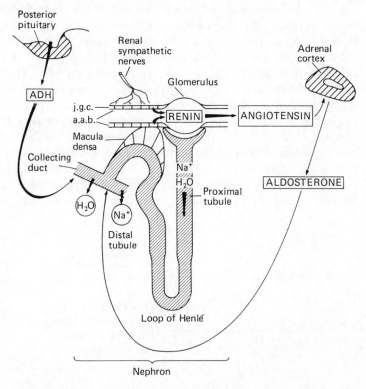

Fig. 2.6. Schematic diagram of a nephron showing some of the important endocrine influences on the kidney. When blood pressure and volume are reduced, renin is released from the juxtaglomerular apparatus. Renin acts on substrate in the plasma to form angiotensin. Angiotensin acts on the adrenal cortex to release aldosterone which increases renal retention of sodium. Depletions of both the cellular and extracellular fluid compartments stimulate the release of antidiuretic hormone (ADH) which increases renal water retention. j.g.c., juxtaglomerular cells; a.a.b., afferent arteriolar baroreceptors.

acts as an endocrine organ, secreting the hormone renin. It will become clear in this book that this endocrine function of the kidney is important for the maintenance of fluid balance.

The kidney contains many nephrons (approximately one million in man), which are long unbranched tubules (see figure 2.6). At the entrance to the nephron is the glomerulus which is highly vascularized. Along the walls of the arterioles leading into the glomerulus are the juxtaglomerular cells which are the principal source of renin. The juxtaglomerular cells and the nearby macula densa together are called the juxtaglomerular apparatus. Renin is released when water or sodium needs to be conserved. For example, if blood volume decreases, a reduction in intra-arteriolar pressure (sensed by the afferent arteriolar baroreceptors) at the level of the juxtaglomerular cells will lead to the release of renin. Renin is also released when there are changes in the sodium concentration of the fluid reaching the macula densa (the nature of the critical sodium chloride change is under dispute). The renal sympathetic nerves are also involved in the control of renin release. Renin acts on a substrate in plasma to form antgioensin I which converting enzyme changes into the octapeptide angiotensin II (see Chapter 4). Angiotensin II has a number of actions which help to restore plasma volume and sodium concentration (see figure 2.7). This restoration of the plasma and the increase in circulating angiotensin act on the juxtaglomerular cells to inhibit further renin release, thus forming a negative-feedback loop. Angiotensin helps to replenish lost

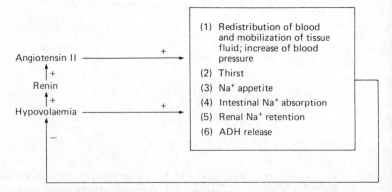

Fig. 2.7. Angiotensin II, the hormone of blood volume regulation. The six actions of angiotensin listed in the figure help to restore the blood volume and pressure in various hypovolaemic states. Some of these actions are direct (e.g. the stimulating action of angiotensin on thirst). Others are indirect (e.g. the action of angiotensin in promoting sodium ion retention by the kidney is mainly mediated by aldosterone). (Modified from Fitzsimons, 1976.)

fluid and electrolytes by stimulating both thirst and sodium appetite (see Chapter 4). The finding that the kidney is involved in fluid intake alters our view that it is an organ primarily for filtering plasma to form urine.

Blood enters the glomerulus under high pressure and a protein-free fluid is driven across the capillary membranes to the long tubules of the nephron. It is in these tubules, which include the proximal tubule, the loop of Henlé, the distal tubule and the collecting duct, that the final volume and composition of the urine are determined by processes which include active transport, absorption, diffusion and osmosis. Details of these processes can be found in any textbook of physiology (e.g. Vander, Sherman & Luciano, 1975; Ganong, 1979). For the understanding of thirst and fluid balance the most important point is that the final changes in the volume and composition of the urine which take place in the distal tubule and collecting duct are under hormonal control. When the cells of the body become dehydrated or the plasma volume falls, these depletions are detected by special receptors in the brain and periphery which control the release of antidiuretic hormone (ADH, also called vasopressin). ADH is formed in cells of the supraoptic and paraventricular nuclei (see figure 9.1) in the hypothalamus and is transported along the hypothalamoneurophypophyseal tract to the posterior lobe of the pituitary where it is stored and released into the blood stream. ADH then acts on the kidneys, increasing the permeability of the collecting ducts, and thereby promoting the reabsorption of water and the formation of a hypertonic urine. When ADH is present, small volumes of concentrated urine are excreted and an excess of water over solute is retained. When ADH is absent, as for example in diabetes insipidus (see Chapter 9), a copious dilute urine is passed and there is a net loss of fluid from the body. Thus, these changes in urine flow influence the entire fluid balance of the organism. Like angiotensin, ADH forms part of a feedback loop controlling fluid balance (see figure 2.8). Unlike angiotensin, ADH has little proven effect on fluid intake (see B. J. Rolls, 1971) although there may be species variations (Kozlowski & Szczepanska–Sadowska, 1975). ADH does, however, affect fluid intake indirectly through its influence on urinary water loss (see discussion of diabetes insipidus in Chapter 9).

Conclusions

Water, which is the main component of the body, is divided between the (intra)cellular and extracellular fluid compartments. These compartments are separated by a semipermeable membrane which

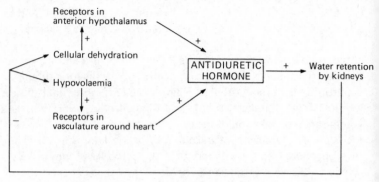

Fig. 2.8. Antidiuretic hormone (ADH) forms part of a feedback loop controlling fluid balance. ADH release is also stimulated by various other factors including stress, pain, exercise and a variety of drugs including morphine, nicotine and barbiturates, and probably angiotensin II. Alcohol decreases ADH release and therefore stimulates diuresis.

allows water and some small ions to pass freely but which prevents movement of sodium and chloride ions and protein. Thus most of the sodium and chloride ions are in the extracellular fluid while most of the potassium is intracellular. The volume and composition of the fluid compartments must be maintained relatively constant. Depletions of both compartments activate mechanisms for conserving fluid (i.e. the release of antidiuretic hormone which helps to prevent renal water loss) and for increasing fluid intake. Manipulations of the fluid compartments form the basis of many studies of thirst. The cells can be dehydrated easily by increasing the sodium concentration of the extracellular fluid. Since sodium does not cross the cell membrane, water is drawn out of the cells by osmosis. Cellular volume is difficult to measure directly, but the degree of cellular dehydration can be inferred from measures of plasma sodium concentration or plasma osmolality. The extracellular compartment can be depleted by removing whole blood or isotonic protein-free fluid (see Chapter 4). Measures of plasma protein or haematocrit give an indication of changes in plasma volume. Decreases in plasma volume, which are a critical thirst stimulus, can be detected by vascular receptors in and around the kidneys. In response to this depletion the kidneys release the enzyme renin which acts on a substrate in the plasma to form angiotensin 1 and II. Angiotensin II stimulates fluid intake. Thus there are several different systems which act to ensure that fluid balance is maintained. This is vital because the body constantly loses water and most terrestrial vertebrates lack a mechanism for storing reserves of water.

3   The origins of thirst

Thirst can be an intense and compelling sensation, yet there is no obvious thirst organ. Subjectively, thirst is closely associated with a dry, tacky, unpleasant-tasting mouth, and some of the oldest theories of the origin of thirst localize the urge to drink in the mouth and throat.

The dry mouth theory

The idea that thirst is related to the relative dryness of the mouth and throat dates back to the ancient Greeks (see Grossman, 1967). In 1764 Haller clearly stated that 'thirst is seated in the tongue, fauces, oesophagus, and stomach'. He attributed this sensation to a deficiency of mucus and saliva. The most influential statement favouring a dry mouth theory was very similar to that of Haller. In 1919 the American physiologist Cannon put forward the following theory:

. . . the salivary glands have, among their functions, that of keeping moist the ancient watercourse; that they, like other tissues, suffer when water is lacking in the body – a lack especially important for them, however, because their secretion is almost wholly water, and that, when these glands fail to provide sufficient fluid to moisten the mouth and throat, the local discomfort and unpleasantness which result constitute the feeling of thirst.

There is no doubt that salivary flow correlates well with the need for water. In 1947 Adolph found that the rate of salivary flow decreases approximately linearly with body water deficit (figure 3.1). The real question which needs to be examined, however, is whether decreased salivation is the primary cause of thirst. First we will review some of the evidence consistent with a dry mouth theory and then some contrary evidence will be considered.

In addition to studies of correlations between dryness of the mouth and thirst, such as those by Adolph, already mentioned, numerous experimenters have examined the effects on thirst of altering oral sensations or blocking them, and found results consistent with the dry mouth theory. For example, applying cocaine to anaesthetize the mouth was found to relieve thirst in both man and dogs (see Wolf, 1958).

Against the local theory, however, were experiments demonstrating that there could be a dissociation between salivary flow and thirst.

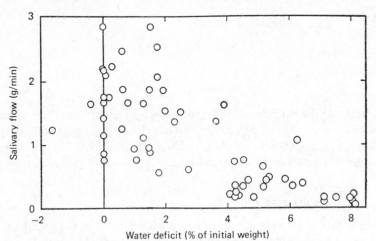

Fig. 3.1. Salivary flow in man decreases approximately linearly with body water deficit. (From Adolph, 1947.)

Removal of the salivary glands is an obvious way to alter salivary flow. It was found that this type of surgical manipulation did not significantly affect water intake in dogs kept at normal room temperature (Montgomery, 1931). Studies of a young man with persistently dry mucous membranes of the mouth due to a total lack of salivary glands from infancy revealed no gross abnormalities of fluid intake. He would relieve the dry mouth every hour with small quantities of water, but reported thirst leading to large drinks about four times a day. His total water intake was not elevated (Steggerda, 1941), but some studies in which the salivary glands were removed suggest that a dry mouth can influence water intake under some conditions. When dogs without salivary glands were forced to pant in the heat (Gregersen & Cannon, 1932) or when salivarectomized rats were fed a dry diet, the resulting dry mouth and throat led to elevated water intake (Epstein, Spector, Samman & Goldblum, 1964) (figure 3.2). These rats did not show global impairments of thirst, for example they drank normal amounts in response to deprivation or osmotic challenges. Various experiments in which the sensations coming from a dry mouth were blocked by neural transection, neural blocking agents, or replications of the earlier local anaesthetic experiments found that thirst persisted even when the dry mouth could not be sensed (see Wolf, 1958; Grossman, 1967).

The most convincing evidence against the dry mouth theory comes from observations of persistent thirst in spite of continual bathing of the mouth and throat with water. In an eighteenth-century children's tale the notorious liar Baron von Munchhausen noticed that his horse was

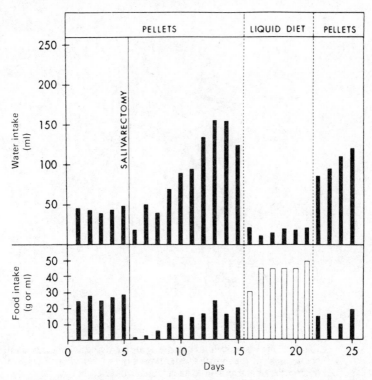

Fig. 3.2. Exaggeration of total daily water intake by the removal of saliva from a rat eating dry food. Daily food and water intake of rat immediately before and after removal of the major salivary glands. (From Epstein, Spector, Samman & Goldblum, 1964.)

unable to quench his thirst. The horse had been cut in two by a portcullis so all the water the animal drank poured out of the other end (see figure 3.3). Liar the Baron may have been but his fantastic story did anticipate the more controlled observations that came the following century. In 1856 Claude Bernard prepared horses and dogs with an oesophageal or gastric fistula (i.e. opening, see figure 3.4). When the animals were tested with open fistulae, they drank enormous quantities of water (sham-drinking), in spite of the mouth being bathed with large quantities of water. In 1939 Bellows confirmed Bernard's early experiments in dogs and observed that during persistent sham-drinking water wets the mouth and therefore the dry mouth theory cannot be correct. Sham drinking had also been seen clinically in 1925 by Gairdner (see Adolph, 1964). In attempting suicide a man cut his oesophagus, but not his arteries or veins. After several days he became intensely

Fig. 3.3. The first (albeit fictitious) suggestion that drinking would proceed continuously if the ingested water flowed out of the animal came with the description of Baron von Munchhausen's bisected horse.

thirsty, but all of the water he drank ran straight out of the oesophagus and produced no relief of thirst. Thirst was, however, relieved by placing water into the lower half of the oesophagus. Recently persistent sham drinking has been observed in rhesus monkeys with surgically prepared gastric or duodenal fistulae (Maddison, Wood *et al.*, 1980). These experiments and the implications for the controls of drinking will be discussed further in Chapters 5 and 6.

Cannon was aware of the studies on sham drinking, but did not allow that they disproved the dry mouth theory. He argued that in the brief pauses in sham drinking the mouth dried out quickly and thereby restarted the drinking. This argument would not convince anyone who has ever observed an animal sham drinking. The animals drink quantities which are considerably in excess of normal levels before any pauses occur, and then the pauses are usually very brief.

This evidence indicates that a dry mouth and thirst can be dissociated, and that therefore there must be some other explanation for thirst. However, the role of dryness of the mouth in thirst should not be

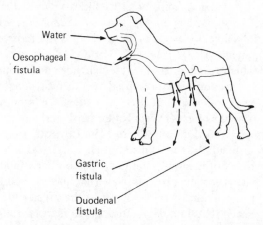

Fig. 3.4. Some of the types of fistulae (openings) that have been used to study sham drinking. The studies were pioneered by Bernard in 1856. When any of the three fistulae are opened the animals drink continuously even though the mouth is continually bathed with fluid, indicating that a dry mouth is not an essential thirst stimulus. When the fistulae are closed the animals function quite normally.

completely disregarded. Previous work (see Wolf, 1958) and our own studies of thirst in humans show clearly that normally the subjective sensation of thirst has a very strong oral component described as a dry, tacky, putrid-tasting mouth (see further Chapters 1 and 5). In severe dehydration the saliva becomes viscous, the mouth feels rough and irritated, the tongue feels swollen and it is difficult to swallow. People living in the desert keep the mouth moistened to decrease thirst by taking small drinks or by exciting salivation by carrying stones or other insoluble substances in the mouth, or by taking into the mouth fruit acids such as lemon or tomato juice. Clinically it is found that moistening the mouth helps to relieve the thirst of patients not allowed fluids.

Thirst of general origin

In the sham-drinking animal hydration of the body through the lower portion of an oesophageal fistula is much more effective in stopping drinking than local hydration of the mouth (Bernard, 1856). This implies that thirst is a more general sensation than local dehydration of the oral cavity and throat. In the nineteenth century it became clear that dehydration of the tissues and/or the blood must be an important factor in thirst. Dupuytren (cited by Rullier, 1821) found that intravenous injections of water or milk or other fluids relieved the thirst of dogs dehydrated by running in the sun. In 1823 Magendie applied this observation clinically by giving intravenous water to relieve the thirst of a rabid patient. A few years later Latta (1832) treated the circulatory collapse which resulted from severe diarrhoea and vomiting in cholera with an intravenous mixture of saline and bicarbonate of soda. This mixture gave immediate relief to the patient's symptoms including the excessive thirst. The experiments of Bernard on sham drinking described in the previous section followed in 1856. It was thus clearly established both experimentally and clinically that thirst was not simply due to local dryness of the mouth, but depended on body fluid changes.

The next important question was what specific bodily changes give rise to the sensation of thirst. Early in the nineteenth century it had been suggested that fluid deprivation might increase the viscosity of the blood. However, since there was no direct evidence confirming this hypothesis, the dry mouth theory was favoured until the early twentieth century when firm evidence was produced that changes in the body fluids could influence drinking. In 1900 Mayer found that after 5 to 9 days of fluid deprivation the osmolality of the blood serum of dogs was elevated. He suggested that hypertonicity of the blood could affect drinking by exciting receptors in the walls of the vascular system which in turn activated receptors in the mouth and throat. Thus he held that the dry mouth is due to systemic changes, but that thirst is translated into drinking behaviour by an awareness of the dryness of the mouth. Mayer's experiments were influential in drawing attention to the importance of extracellular osmolality in thirst and were quickly followed by those of Wettendorff in 1901 who also found osmotic changes of the blood produced by water deprivation and was the first to suggest that cellular dehydration is a thirst stimulus. As we will see in Chapter 4, this notion has been confirmed by a range of experiments. Wettendorff also made the important observation that thirst follows haemorrhage. Not all of Wettendorff's suggestions showed such brilliant insight. He proposed that all cells of the body contribute to the sensation of thirst or, in other words, that every cell is a thirst receptor which sends signals to a central co-ordinator responsible for the conscious awareness of thirst. Much experimental evidence refutes such a general notion of thirst reception, and even before Wettendorff had published his ideas there was evidence that thirst receptors are localized in particular areas, especially in the central nervous system (see Chapter 4).

Early studies of the central nervous system and thirst

The theories of Mayer and Wettendorff required the involvement of the peripheral nervous system to detect peripheral changes in fluid balance. Wettendorff had only the vague notion that the brain is somehow responsible for the conscious awareness of thirst. Mayer was more specific in proposing a brainstem centre which regulates drinking. The notion of the localization of specific functions to specific brain areas was well established by the end of the nineteenth century and the idea that thirst might have a specific regulatory centre in the brain seemed to be supported by several clinical reports. In 1881 Nothnagel reported that a man who had been kicked by a horse, and in falling received a blow to the back of the head, became excessively thirsty half an hour later. Over the next 3 h he drank 3 l of fluid, but did not urinate until 2.5 h after the start of drinking. The excessive thirst persisted over the next 2.5 weeks of observation. Nothnagel indicated that his observation supported the notion of a specific thirst centre in the brain, probably at the back of the brain in the medulla or pons. More clinical reports of excessive thirst followed that of Nothnagel. However, such clinical observations, although very suggestive of a central thirst centre and invaluable in drawing attention to particular brain areas, are not definite proof for the central control of drinking. This is because the drinking could be secondary to increased urine flow (due to damage to the neural control of antidiuretic hormone release), and this was not always tested. Also, autopsies are rarely carried out to determine the exact site in the brain where damage has occurred.

Later experiments demonstrating that antidiuretic hormone release is controlled by specific brain areas were instrumental in directing experimental attention to manipulations of possible thirst centres. In 1947 Verney published his classical review of his experiments on the control of antidiuretic hormone. He manipulated the concentration of the blood flowing into the brain by infusing hypertonic sodium chloride solutions into the carotid arteries of dogs. He found that increasing the osmolality, and thereby drawing water out of cells in the brain, led to the release of ADH (figure 3.5). By combining carotid injections with ligation of specific parts of the blood supply to the brain, Jewell & Verney (1953, 1957) were able to suggest that specific receptors which responded to the osmolality of the blood by releasing ADH, and thereby conserving fluid, were localized in the region of the hypothalamus, possibly in or near the supraoptic nucleus. Shortly after this elegant demonstration that there are central 'osmoreceptors', Wolf (1950) proposed that similar receptors are involved in the control of drinking.

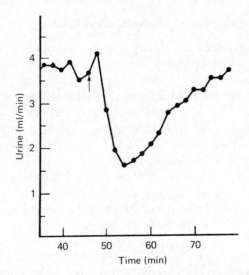

Fig. 3.5. An example oɪ a classical experiment by Verney which showed for the first time that there are osmoreceptors in the brain which respond to dehydration produced by hypertonic sodium chloride by stimulating the release of ADH. At the arrow 20 ml of 0.257 M sodium chloride was injected into the left carotid artery over 16 s. The subsequent rise in ADH reduced urine flow. (From Verney, 1947.)

By this time the stereotaxic instrument which permitted stimulation of or damage to specific brain areas was in wide use. In 1952 Andersson was able to demonstrate that injections of hypertonic sodium chloride directly into the hypothalamic region of goats elicited copious drinking after only a short delay. This was a classical experiment in the history of thirst in that it led to numerous experiments searching for the central controls of drinking. Some of these experiments localizing central osmoreceptors are discussed in Chapter 4. Chapter 7 is devoted entirely to the neural control of drinking.

This chapter outlines some of the important landmarks in the early history of the study of thirst. The 'modern era' is introduced by topic in the later chapters. For those interested in delving further into the historical antecedents of current thinking about thirst excellent reviews can be found in the following sources: Wolf (1958), Grossman (1967), and Fitzsimons (1973, 1979).

Conclusions

The notion that thirst is due to local dryness of the mouth and throat is supported by subjective reports of dehydrated people, but many experiments show that thirst must also be due to more general body fluid changes. Studies of sham drinking in which water bathes the mouth and oesophagus, but is drained out before being absorbed, indicate that thirst persists when the mouth and throat are being moistened continually. Early in the nineteenth century it was found that intravenous water or saline gave relief to thirst. Thus it was suggested that thirst is due to general body fluid changes. Not until the beginning of the twentieth century did the specific fluid changes responsible for thirst start to be understood. Studies of the osmotic changes in the blood which occur during dehydration led to the suggestion that loss of cellular water is a thirst stimulus. At first it was thought that all cells of the body could detect such dehydration, but it was later thought more likely that thirst receptors were localized in specific regions of the brain. Confirmation of this theory came from studies which showed that injections of hypertonic salt solutions into the hypothalamus caused copious drinking. The search for a precise characterization of the factors responsible for the initiation, maintenance and termination of drinking and the neurology of thirst forms the basis for much of the recent work on thirst and is described in detail in the following chapters.

4   The initiation of drinking

Cellular dehydration as a thirst stimulus

At the beginning of this century Mayer and Wettendorff (see Chapter 3) showed that thirst could be stimulated by dehydration of the body fluids. The critical fluid changes which influence drinking have only been defined recently and although this is the area where most work on thirst has been concentrated there is still much scope for further experimentation. This chapter is devoted to the fluid changes involved in the initiation of drinking. In Chapter 6 the fluid changes associated with the termination of drinking are discussed.

We have seen that the body fluids are divided between the cellular and extracellular compartments. Defining the critical thirst stimuli has involved manipulation and measurement of both compartments and relating changes in these compartments to drinking behaviour. The cellular compartment will be considered first and then the extracellular compartment. The effects of changes in both compartments will be considered in the discussion of thirst induced by water deprivation.

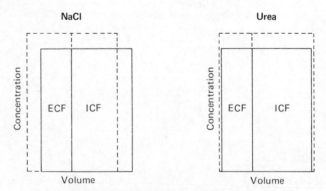

Fig. 4.1. Effects of administration of hypertonic sodium chloride and equiosmotic urea on body fluid distribution. The solid lines represent normal fluid balance and the dotted lines indicate the equilibrium achieved at the end of the infusions. The cell membrane is impermeable to sodium chloride which stays in the extracellular fluid and dehydrates the cells by osmosis. Urea, which crosses the membrane and distributes between the cellular (ICF) and extracellular (ECF) fluid compartments, does not affect cell size.

In 1937 Gilman demonstrated that a critical stimulus for drinking following dehydration is loss of water from inside the cells. His simple yet elegant demonstration that cellular dehydration and not absolute osmotic pressure is the stimulus for drinking is shown diagrammatically in figure 4.1. He found that the administration of hypertonic solutions of substances to which the cell membrane is impermeable, such as sodium chloride or sucrose, stimulates drinking because the substances remain outside the cells and thus cause a loss of water from inside the cells by osmosis. On the other hand, the administration of similar concentrations of substances such as glucose, urea, and methyl glucose, which cross the cell membrane and do not therefore lead to cellular dehydration, stimulates little or no drinking (Gilman, 1937; Fitzsimons, 1961a). The degree of cellular dehydration is monitored accurately in that there is a precise relation between the amount of sodium chloride administered and the amount drunk in nephrectomized rats (i.e. rats whose kidneys have been removed surgically) in which the stimulus cannot be modified by urine production (see figure 4.2).

Recently, the cellular dehydration theory has been challenged by Andersson & Olsson (Andersson & Olsson, 1973; Andersson, 1978). They suggest that the critical stimulus may be a change in the sodium concentration of the cerebrospinal fluid (CSF). They agree that the receptors for cellular thirst are located in the central nervous system

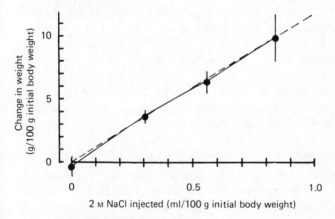

Fig. 4.2. The net fluid intake (i.e. change in body weight) by nephrectomized rats 6 h after injection of hypertonic sodium chloride. The broken line indicates the calculated changes in body weight required to dilute the hypertonic injections to isotonicity on the assumption that no excretion occurs. Unless otherwise stated, all figures throughout the book show means and standard errors of the means. (After Fitzsimons, 1966.)

but propose that they respond to changes in the sodium concentration of the CSF rather than to cellular dehydration. A critical test of this theory is to apply an osmotically active substance other than sodium to the receptors in the brain and to determine whether drinking follows. Blass & Epstein (1971) and Peck & Novin (1971) found that the direct application of sucrose to central osmoreceptors via cannulae implanted stereotaxically stimulates drinking in the rat and rabbit. In a different type of test the sodium content of the cerebrospinal fluid was measured and manipulated. It was found that drinking correlates poorly with the measured concentration of sodium in the cerebrospinal fluid (Epstein, 1978). Instead drinking was related to withdrawal of water from the cells (i.e. the effective osmotic pressure) during ventricular infusions of sodium or sucrose (Thrasher, Jones *et al.*, 1980; Ramsay, Thrasher & Keil, 1980). So we have come full circle back to Gilman's suggestion that dehydration of cells is a crucial stimulus of thirst and that this, rather than sodium concentration or absolute osmotic pressure is sensed.* It will become clear throughout this chapter that cellular dehydration accounts for a large portion of normal drinking.

Before proceeding to the question of where the receptors for cellular thirst are located, it is essential to clarify one point. We have just reviewed the evidence which indicates that cellular dehydration and not general increases in osmolality or sodium concentration constitute the crucial thirst stimulus. In practice, however, plasma sodium concentration and plasma osmolality are highly correlated (Wood, Rolls & Ramsay, 1977) and both give a good indication of the degree of cellular dehydration. This is because sodium and its attendant anions account for approximately 90–95% of plasma osmolality so that changes in sodium concentration acting through cellular dehydration will be the most important factor in cellular thirst. In the past the cells which respond to cellular dehydration have been referred to as 'osmoreceptors' and we will continue this convention.

Localization of the osmoreceptors

It is thought that there are specific osmoreceptors for thirst similar to those described by Verney (1947) for the release of antidiuretic hormone. Various possible locations for such thirst receptors have been suggested, including the stomach, gut, hepatic–portal circulation and the mast

* Andersson and Olsson studied sodium receptors in the goat. Recent Australian work (McKinley, Denton, Leksell *et al.*, 1980) suggests that there may be both sodium and osmoreceptors in the sheep. It thus remains possible that ruminants which, unlike the dog and rat, have a large turnover of sodium may have both sodium receptors and osmoreceptors.

cells in the peritoneal cavity. Since 1953, when Andersson found that injections of hypertonic sodium chloride solutions directly into the hypothalamus caused drinking in the goat, it has been presumed that there are osmoreceptors for thirst in the central nervous system. Andersson's doses of saline were, however, large and outside the physiological range so that the thirst could have been due to non-specific stimulation.

One way of establishing clearly whether there are receptors in the brain that respond to dehydration is to increase the tonicity of the blood perfusing the brain. This technique provides a more physiological stimulus than does direct application of substances to the brain and has been utilized by Wood, Rolls & Ramsay (1977). The main blood supply to the forebrain is provided by the carotid arteries, and in a large animal such as the dog carotid loops can be prepared. This is a simple surgical procedure in which the arteries are exteriorized and then the loose skin of the neck sewn around them.

The arteries are thus readily accessible for injections or infusions. An

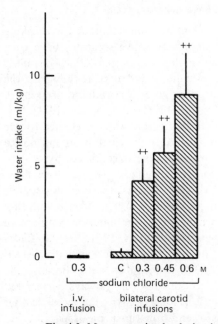

Fig. 4.3. Mean water intake during infusions of hypertonic NaCl in eight fluid-replete dogs. All carotid infusions were bilateral at a total rate of 0.6 ml kg^{-1} min^{-1} for 10 min and water intake was measured for the last 5 min. Control infusion (C) was 0.15 M sodium chloride. i.v., intravenous. ++, significantly different from control. (Modified from Wood, Rolls & Ramsay, 1977.)

advantage of this technique is that the osmolality of the blood perfusing the brain can be altered with little change in the osmolality of the rest of the body. Dogs which had been prepared with bilateral carotid loops were infused for 10 min with a hypertonic sodium chloride solution in doses designed to elevate central osmolality within the physiological range. A graded increase in drinking was obtained with increasing concentrations of the saline (see figure 4.3). It was shown by analysis of plasma samples taken from the jugular vein that the infusions which elicited drinking did elevate central osmolality within the range which can occur normally after, for example, water deprivation. It was also shown that these small carotid infusions had no significant effect on peripheral systemic osmolality. Control intravenous (as opposed to intracarotid) infusions of 0.3 M sodium chloride at the same rate did not cause drinking.

Thus these results show clearly that there are receptors for thirst in the brain and that they are located within the area supplied with blood from the carotid arteries.

Osmoreceptors in the central nervous system

Further evidence that there are central osmoreceptors for drinking comes from studies in which osmotically active solutions have been injected intracranially. Relatively small volumes (0.5–2.0 μl) of mildly hypertonic (0.18–0.30 M or more) saline injected bilaterally into the lateral preoptic area (for anatomy see figure 7.1) produced drinking in the rat (Blass & Epstein, 1971) and rabbit (Peck & Novin, 1971). The effect of the injections is specific in that eating was not elicited by the injections. The drinking is ascribed to cellular dehydration and not to activation of a sodium receptor in that similar injections of hypertonic sucrose also elicited drinking. Further, equimolar injections of urea, which do not cause cellular dehydration because the urea crosses the cell membrane, did not elicit drinking (figure 4.4). The effective injection sites were dispersed relatively widely, from the anterior commissure anteriorly to the zona incerta posteriorly (Blass, 1974; Peck & Blass, 1975), so that the osmoreceptors themselves could be widely distributed in the lateral preoptic and lateral hypothalamic areas. Because the thirst-inducing action of these injections does not always correspond with their effect on antidiuretic hormone release, it is possible that osmoreceptors for ADH release and for thirst are separate, but contiguous (Peck & Blass, 1975). Thus this evidence suggests that there are osmoreceptors in the brain with a rather widespread distribution in the lateral preoptic and lateral hypothalamic areas whose activation can lead to drinking.

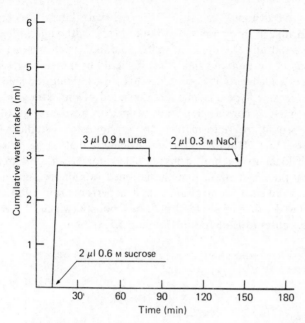

Fig. 4.4. Specificity of drinking in response to hypertonic solutions injected into the lateral preoptic area. Hypertonic NaCl and sucrose elicited drinking. Hyperosmolal urea did not. (From Blass & Epstein, 1971.)

Apparently consistent with the view that there are osmoreceptors for drinking in the lateral preoptic area is the report that bilateral lesions of the lateral preoptic area abolish drinking in response to systemic loads of sodium chloride in the rat (Blass & Epstein, 1971) and rabbit (Peck & Novin, 1971). However, the effect of the lesions is complex in that in some animals there is a specific deficit to cellular dehydration alone whereas in others there is a failure to drink following depletion of either the cellular or the extracellular fluid compartment (Blass & Epstein, 1971; B. J. Rolls, 1975). A further complication is that rats with lesions in the preoptic area, if given enough time after an intraperitoneal injection of sodium chloride, drink almost exactly what they need for osmoregulation (Coburn & Stricker, 1978), as if they could respond to osmotic stimuli. A similar conclusion was reached independently by B. J. Rolls who observed that intraperitoneal injections of 2 M sodium chloride reduced the activity of the lesioned rats, as if they were particularly sensitive to the stress of this procedure. She therefore measured drinking after subcutaneous injections of a smaller osmotic load (1 M sodium chloride) over a longer period (6 h) and found that the lesioned rats did drink in response to the osmotic load, but that the drinking was slow. Thus rats with lateral preoptic lesions can drink in response to osmotic loads, although a long observation period and the use of a saline solution which is not too hypertonic may be necessary to show this. It should be noted, however, that in these extended tests it is possible that, in excreting the solute load, the rats became hypovolaemic (i.e. depletion of the extracellular fluid compartment occurred) and thus drank in response to this stimulus. It would be informative to repeat these studies in nephrectomized rats. However, further evidence that large intra-peritoneal salt loads may inhibit drinking behaviour because of stress is provided by the observation that the lesioned rats drank amounts similar to those taken by control animals if hypertonic salt solutions were administered either in the food or intravenously (which, if given at a slow rate, causes little discomfort) (figure 4.5) (Coburn & Stricker, 1978).

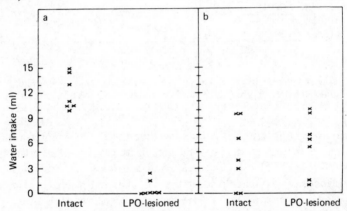

Fig. 4.5. Water intake of seven intact rats and seven rats with lesions in the lateral preoptic area (LPO-lesioned) during a 4-h drinking test following the administration of (a) 2 ml of 2 M sodium chloride solution intra-peritoneally or (b) 4 ml of 1 M sodium chloride solution intravenously. (From Coburn & Stricker, 1978.)

Thus the lesion evidence does not provide useful support for the hypothesis that osmoreceptors for drinking are in the lateral preoptic area. Lesions more medial, in the periventricular tissue at preoptic and hypothalamic levels, particularly in the anteroventral part of the third ventricle (AV3V), can reduce drinking in response to osmotic as well as some other thirst challenges (Buggy & Johnson, 1977a, b) but the reduced responses are not fully understood as yet. Consideration of the function of this region of the brain in drinking is left until later (Chapter 7).

The hypothesis that cells in the preoptic region have a function in responses to osmotic stimuli receives some support from electrophysiological investigations. It has been shown in a number of studies that (large) infusions (in most studies intracarotid infusions) of hypertonic saline influence the activity of some neurons in the preoptic area and hypothalamus, as well as in the supraoptic nucleus, of the rat (Cross & Green, 1959; Blank & Wayner, 1975; Malmo & Mundl, 1975; Weiss & Almli, 1975) the monkey (Hayward & Vincent, 1970; Vincent, Arnauld & Bioulac, 1972; see also Hayward, 1977), and the cat (Emmers, 1973). Of the responsive neurons, some in the supraoptic nucleus itself were activated antidromically from the posterior pituitary and are neurosecretory cells for antidiuretic hormone, and others close to the supraoptic nucleus which did not respond to other non-osmotic stimuli could be osmoreceptors in synaptic contact with neuroendocrine cells (Hayward & Vincent, 1970; Vincent, Arnauld & Bioulac, 1972; Arnauld, Dufy & Vincent, 1975 – for anatomy see figure 7.1). Other neurons scattered through the lateral preoptic region and hypothalamus which responded to the hypertonic infusions could be involved in drinking, but there is no proof of this. Whether these cells responded directly to the hypertonic saline (i.e. were osmoreceptors) or were instead influenced by other osmoreceptor cells is not known, but Oomura, Ono, Ooyama & Wayner (1969) did suggest that some neurons in the lateral hypothalamus which responded to the iontophoretic ejection of sodium ions were osmoreceptors.

Thus there is electrophysiological evidence that neurons in the preoptic area and hypothalamus could be involved in responses to osmotic changes, but whether their activation is involved in drinking is not yet clear.

Peripheral osmoreceptors

As shown by intracranial injections and intracarotid infusions of solutions, the osmoreceptors in the central nervous system respond directly to changes in the tonicity of cerebral blood. It is possible that peripheral thirst receptors also feed information about the state of hydration of the rest of the body into the brain. Receptors which respond to changes in osmolality are widely distributed in the stomach, gut (Hunt, 1956), and hepatic–portal system (i.e. the blood vascular system carrying absorbed substances from the gut to the liver; see figure 6.1). It has even been suggested that mast cells (wandering cells containing heparin and histamine) which are abundant around small blood vessels, might act as receptors that respond to cellular dehydration and induce drinking (Goldstein & Halperin, 1977). Of the possible peripheral osmoreceptors those in the hepatic–portal system have received the most attention. Such receptors were originally studied in relation to the secretion of antidiuretic hormone (Haberich, 1968), but there are now studies on both the dog and the rat which suggest that hepatic–portal osmoreceptors may be involved in the control of drinking. In the dog infusions of water into the hepatic–portal vein elevate the threshold for drinking in response to cellular dehydration produced by systemic hypertonic sodium chloride infusions (Kozłowski & Drzewiecki, 1973). Information about peripheral hydration appears to travel to the central nervous system via the vagus nerve since vagotomy abolishes the effect of these hepatic infusions on drinking in the dog (Kozłowski & Drzewiecki, 1973), and disrupts drinking in response to a hypertonic challenge in the rat (Kraly, Gibbs & Smith, 1975). It has also been shown electrophysiologically that perfusion of the portal vein with solutions which alter the osmolality of the blood reaching the liver causes changes in the discharge of the hepatic vagus nerve (Adachi, Niijima & Jacobs, 1976) and of certain lateral hypothalamic neurons (Schmitt, 1973).

To investigate the role of peripheral osmoreceptors in drinking, we determined whether increased peripheral osmolality can stimulate drinking when the blood perfusing central osmoreceptors remains isotonic. Water intake was measured in dogs while an intravenous infusion of hypertonic sodium chloride which elevated both systemic and central osmolality (to a degree which occurs after water deprivation) was combined with an intracarotid infusion of water, which kept the blood perfusing the brain isotonic. With such selective peripheral hyperosmolality, the dogs did not drink. Thus peripheral hyperosmolality alone at levels which occur after water deprivation is not a sufficient stimulus for drinking (Wood, Rolls & Ramsay, 1977). This conclusion is further supported by our experiment in which attenuation of central hyperosmolality by carotid water infusions markedly reduced drinking by water-deprived dogs despite persisting peripheral hyperosmolality. Thus we conclude that information from periphera, osmoreceptors is not necessary, and is not sufficient when acting alonel for the initiation of drinking in response to physiological levels of dehydration. If peripheral osmoreceptors are involved in the initiation of drinking, their effects must be dependent on the state of central osmoreceptors. The role of peripheral osmoreceptors in the termination of drinking is discussed in Chapter 6.

Summary – cellular dehydration as a thirst stimulus

A critical stimulus for drinking is loss of water from inside the cells, i.e.

cellular dehydration. By convention the receptors for this stimulus are called osmoreceptors. Intracarotid infusions which elevate the effective osmotic pressure of the blood in the head but have no significant effect peripherally stimulate drinking. Thus there are osmoreceptors in the brain. Many different sites have been suggested for the central osmoreceptors, but the areas which have had the most experimental attention are the preoptic area and the region near the anteroventral part of the third ventricle. Lesions in these areas reduce drinking in response to hypertonic thirst stimuli and local injections of physiological levels of hypertonic saline or sucrose stimulate drinking. Further work is necessary to determine whether thirst osmoreceptors are localized in one or two discrete brain regions or whether they are widely distributed in the central nervous system. It should be pointed out that a thirst osmoreceptor has not been identified or characterized histologically. Osmoreceptors also appear to be fairly widely dispersed peripherally, for example in the gut, the hepatic–portal system and in the mast cells. The significance of these peripheral receptors for normal drinking and the way in which they interact with central receptors is still not understood, but it is clear from intracarotid infusion experiments that peripheral increases in effective osmotic pressure, when there are no central changes, are not sufficient alone to stimulate drinking.

Extracellular fluid depletion as a thirst stimulus

Thus far we have considered only the effect of loss of fluid from inside the cells on thirst. While there is no doubt that cellular dehydration is an important thirst stimulus, there is evidence that drinking can be influenced by other fluid changes. Although the amount of fluid in the extracellular compartment is less than that in the cells, it is vital that the extracellular fluid balance be rigorously maintained to avoid debilitating changes in the vascular fluid volume and pressure which, if lowered, could lead to circulatory collapse. As pointed out in Chapter 2, there is a large turnover of fluid in the body. The exchange of water between the body and the environment is effected through the extracellular fluid. A variety of controls exist to maintain the stability of the extracellular fluid in the face of this rapid turnover of water. For example, the role of the kidneys is discussed in Chapter 2.

The extracellular compartment is divided into two components, the intravascular fluid which contains plasma and the extravascular or interstitial fluid. These two phases of the extracellular compartment are usually in equilibrium, the steady state between them being maintained by the exchange of water and ions across the capillary walls. Most of the controls for maintaining fluid balance are located inside the vasculature, which consists of the high-pressure arterial system and the low-pressure venous system. Inside the vasculature are stretch receptors and pressure receptors (baroreceptors) which, when plasma volume is decreased, stimulate the release of antidiuretic hormone and the renal conservation of water. There may also be a need to replenish lost fluid and this must be done by increasing fluid intake. The evidence that depletion of extracellular fluid volume, specifically plasma volume (hypovolaemia), may be involved in thirst will now be considered.

Clinically, haemorrhage is found to lead to drinking (Wettendorff, 1901). The crucial point here is that haemorrhage represents a loss of isotonic fluid which selectively depletes the extracellular fluid compartment, with no osmotic effects. The patient's thirst is found to be relieved by saline injections, which replete the extracellular fluid compartment.

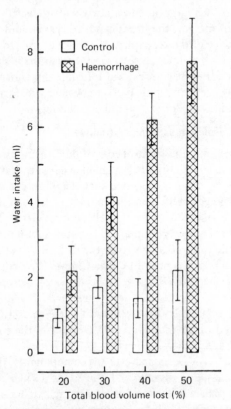

Fig. 4.6. Total water intake over 5 h after graded levels of haemorrhage in the rat, compared with intake in control animals. (From Russell, Abdelaal & Mogenson, 1975.)

Haemorrhage also stimulates drinking in the rat (Fitzsimons, 1961b), and the drinking is related to the volume of blood lost (figure 4.6). Haemorrhage is not, however, an ideal way to manipulate extracellular fluid volume since the removal of the red cells can lead to debilitating anaemia, and plasma volume can be restored by movement of interstitial fluid into the vascular compartment.

Another way of decreasing the extracellular fluid volume is to deplete the body of sodium by lowering the sodium content of the diet. As the sodium concentration decreases there is a loss of extracellular water after a delay, and it is found that the cells may become overhydrated. In spite of this cellular overhydration, the extracellular volume depletion leads to increased fluid intake in a variety of species. In man sodium depletion leads to increased water intake (McCance, 1936), though the subjects sometimes described the subjective sensation as aberrant taste rather than thirst. Rats, dogs and rabbits also drink more water after sodium depletion (see Fitzsimons, 1971).

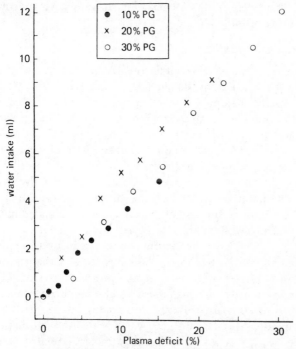

Fig. 4.7. There is a highly significant correlation ($r = +0.98$) between the estimated plasma deficits and the observed water intake of rats injected subcutaneously with 5 ml of 10%, 20%, or 30% polyethylene glycol (PG) solutions. Each point represents the mean value from 11 rats. (From Stricker, 1968.)

Extracellular fluid can be removed quickly and simply by injecting high concentrations of colloids (i.e. glue-like substances made up of large molecules that cannot cross cell membranes; examples are gum acacia or polyethylene glycol) either into the peritoneal cavity or subcutaneously. Isotonic fluid accumulates around the colloid, thereby depleting the extracellular fluid of both water and sodium. Within 1 h of the injection urine flow decreases and water intake increases. The amount of water consumed depends on the plasma deficit (figure 4.7) (Fitzsimons, 1961b; Stricker, 1968). A marked appetite for sodium also develops, and the subsequent ingestion of sodium restores the extracellular volume and composition to normal (Fitzsimons, 1971; Stricker, 1981).

Thus, changes in the extracellular fluid compartment can stimulate drinking when there are no changes in or even when there is over-hydration of the cellular volume. The critical depletion is probably reduced plasma volume or hypovolaemia. The final repletion of the extracellular compartment occurs when sodium appetite develops some time after the onset of thirst.

Angiotensin as a thirst stimulus

The drinking which follows extracellular depletion may be mediated by receptors in the vasculature. Pursuing this idea, Fitzsimons (1964, 1969) experimentally manipulated the circulation without affecting fluid balance. He found that diminishing venous return to the heart and reducing arterial blood pressure by ligation of the inferior vena cava of the rat led to a marked increase of water intake and a positive fluid balance due to decreased urine flow (figure 4.8). To check the possibility that the decreased urine flow after caval ligation might be influencing water intake, Fitzsimons repeated his experiment in a group of rats with the kidneys removed. Surprisingly, caval ligation stimulated much less drinking in the rats without kidneys. This suggested that an essential thirst stimulus following caval ligation might be a reduction in blood pressure to the kidneys. To test this, Fitzsimons reduced the pressure to the kidneys without affecting overall pressure, by partially constricting the renal arteries. This procedure also stimulated copious water intake. From these experiments, Fitzsimons inferred that the kidney could contain a specific substance which is involved in drinking. There had been earlier reports that renal extracts stimulated water intake and Fitzsimons went on to show that this substance was renin (see Fitzsimons, 1969). Renin is an enzyme which acts on substrate in the plasma to form angiotensin I, which is converted to angiotensin II, a vasoactive octapeptide. Fitzsimons and B. J. Rolls showed that the

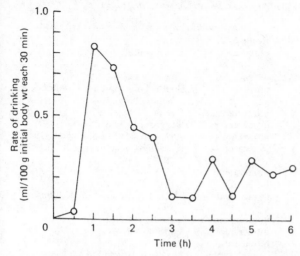

Fig. 4.8. The rate of drinking in the 6 h following caval ligation above the renal veins in the rat. Each point is the mean of 10 observations. (After Fitzsimons, 1969.)

active dipsogen was probably angiotensin II since intravenous infusions of that substance stimulated copious drinking (Fitzsimons & Simons, 1969) (figure 4.9). This then raised the question of how angiotensin elicits drinking. The hypothesis tested by Epstein, Fitzsimons & Rolls (1970) was that it acts directly on receptors in the central nervous system. We found that the direct application of angiotensin to the brain caused rats in water balance to drink large quantities of water. This drinking is unlikely to be due to the spread of the hormone peripherally since the doses which were effective centrally were at least 1000 times smaller than those required peripherally.

The effect of angiotensin is very specific. Drinking is the only behavioural response which follows its administration. For example, after intracranial angiotensin, a sleeping rat woke and went immediately to water. Also, a rat which had been deprived of food, but not of water, stopped feeding to drink (Epstein, Fitzsimons & Rolls, 1970; McFarland & Rolls, 1972; B. J. Rolls & McFarland, 1973). The drinking following angiotensin administration is highly motivated. After intracranial angiotensin rats pressed a lever as many as 64 times for a single reward of 0.1 ml of water (B. J. Rolls, Jones & Fallows, 1972). A wide variety of species tested with angiotensin have shown a copious drinking response (table 4.1). It is unlikely that angiotensin stimulates drinking by non-specific activation of the brain since the

Table 4.1. *Species which drink in response to angiotensin*

Species	Route of administration	Investigators
Rat	intravenous	Fitzsimons & Simons, 1968
	intracranial	Booth, 1968; Epstein, Fitzsimons & Simons, 1969
Cat	intravenous	Cooling & Day, 1974
	intracranial	Sturgeon, Brophy & Levitt, 1973
Dog	intravenous	Trippodo, McCaa & Guyton, 1976
	intracarotid	Fitzsimons, Kucharczyk & Richards, 1978
	intracranial	B. J. Rolls & Ramsay, 1975
Rhesus monkey	intracranial	Setler, 1971
Cebus monkey	intracranial	Block, cited in Schwob & Johnson, 1977
Baboon	intracranial	Lotter *et al.*, 1980
Goat	intracranial	Andersson & Westbye, 1970
Sheep	intracarotid, intracranial	Abraham, Baker, Blaine, Denton & McKinley, 1975
Pig	intracranial	Baldwin, 1979, personal communication
Mongolian gerbil	intracranial	Block, Vallier & Glickman, 1974
Rabbit	intracranial	Findlay, Fitzsimons & Setler, cited in Fitzsimons, 1972
Marsupial possum (Australian)	intravenous	Young & McDonald, 1978
Marsupial opposum (North American)	intracranial, intravenous	Elfont, Epstein & Findlay, 1980
Barbary dove	intracranial	McFarland & Rolls, cited in Fitzsimons, 1972
Pigeon	intraperitoneal, intravenous, intracranial	Evered & Fitzsimons, 1976
White-crowned sparrow	intravenous, intracranial	Wada, Kobayashi & Farner, 1975
Chicken	intravenous, intracranial	Snapir, Robinson & Godschalk, 1976
Japanese quail	intravenous, intracranial	Takei, 1977
Peking duck	intracranial	De Caro, Mariotti, Massi & Micossi, 1980
Eel	intra-arterial	Hirano, Takei & Kobayashi, 1978
Iguana	intraperitoneal	Fitzsimons & Kaufman, 1977
	intracranial	Fitzsimons, 1979
Euryhaline killifish	intraperitoneal	Malvin, Schiff & Eiger, 1980

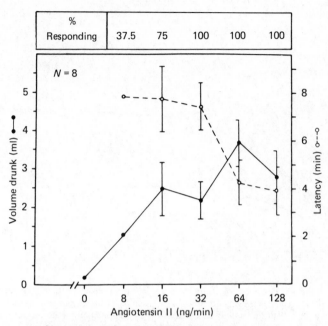

Fig. 4.9. Dose–response relationship between intravenous angiotensin II and elicitation of drinking in water-replete rats. The percentage of rats drinking in response to each dose is given in the boxed heading. (From Epstein & Hsaio, 1975.)

doses used are very small (see below) and other similar vasoactive peptides (e.g. vasopressin) are relatively ineffective.

The effectiveness of angiotensin as a dipsogen is illustrated by the copious drinking it stimulates in animals in fluid balance. Normally endogenous increases in angiotensin would occur along with fluid depletions and it has been found that angiotensin-induced drinking will add to drinking stimulated by cellular dehydration in the rat (Fitzsimons & Simons, 1969). The interactions of thirst stimuli may vary from species to species. For example, in the dog a subthreshold dose of angiotensin will lower the thirst threshold to hypertonic saline (Kozłowski, Drzewiecki & Zurawski, 1972), and in the goat angiotensin and hypertonic saline together elicit more drinking than the sum of the drinking responses to the two stimuli given separately (Andersson & Eriksson, 1971). If rats are overhydrated, a given dose of angiotensin is relatively ineffective (figure 4.10) (B. J. Rolls & McFarland, 1973).

The facts that angiotensin stimulates drinking in a wide variety of species, that it produces highly motivated drinking, and is influenced

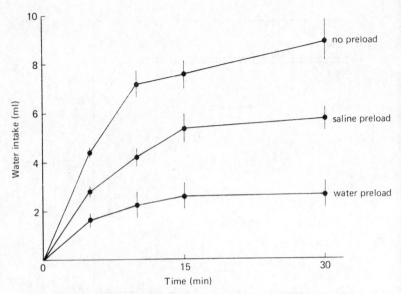

Fig. 4.10. The effect of 10 ng intracranial angiotensin on the water intake of rats when there was no preload or when a 15 ml intragastric preload of water or isotonic saline was administered 15 min before the intracranial injection. Although cellular overhydration with water was most effective in suppressing drinking, isotonic saline preloads also significantly reduced intake. This could be due to ECF expansion or to saline remaining in the gut. The results are shown as means $\pm$S.E. (Modified from B. J. Rolls & McFarland, 1973.)

by the state of the body fluids indicate that angiotensin could play a role in normal drinking. We will return to this issue later (see page 52).

Localization of the receptors for angiotensin

Since the initial discovery that intracranial angiotensin stimulates drinking, much work has been aimed at localizing precisely the receptive site(s). In their original work Epstein, Fitzsimons & Rolls (1970) found the preoptic area to be the most sensitive site, but since angiotensin does not normally penetrate the blood–brain barrier* it was difficult to see how the substance would reach its receptors unless the barrier in this region was permeable to angiotensin. A possible answer to the

---

* The 'blood–brain barrier' refers to the fact that not all substances pass with ease from the blood into the brain. The blood–brain barrier is not uniform throughout the nervous system. In some parts of the brain, such as the circumventricular organs, substances pass from the blood relatively freely. One way in which the blood–brain barrier can be by-passed experimentally is to inject substances directly into the cerebrospinal fluid which fills the ventricles and which is on the brain side of the barrier.

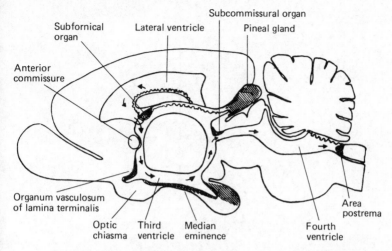

Fig. 4.11. The circumventricular organs (shaded) shown in a diagrammatic saggital section of an adult rat brain. (From Phillips, 1978.)

problem of how angiotensin could reach its receptors is provided by experiments on the circumventricular organs (figures 4.11 and 7.1). These organs on the surface of the ventricles are part of the brain, but lie outside the blood–brain barrier (i.e. the capillaries which supply these organs are much more permeable than those that supply the rest of the brain). The importance of the ventricles for the dipsogenic response was indicated by the work of Johnson & Epstein (1975) which showed that intracranial angiotensin was more effective in stimulating drinking if the cannula for its injection passed through a ventricle. Several circumventricular organs have since been suggested as receptive sites for angiotensin. Local application of angiotensin to the subfornical organ (SFO) stimulated drinking and the threshold dose was very low – between 10^{-15} and 10^{-14} moles (10^{-12} and 10^{-11} g) (see figure 4.12). Furthermore, lesions of the SFO or local application of a competitive angiotensin inhibitor abolished drinking in response to intravenous angiotensin (Simpson & Routtenberg, 1973; Simpson, Epstein & Camardo, 1977). Electrophysiological studies have identified angiotensin-sensitive cells in the SFO (figure 4.14 below) (Felix & Akert, 1974; Phillips & Felix, 1976; S. Nicolaïdis, personal communication).

A controversial issue which has been raised by the work on the SFO is whether this site is unique in containing cells sensitive to the dipsogenic action of angiotensin. Lesions of the SFO completely blocked drinking induced by intravenous angiotensin. However, rats (Hoffman & Phillips, 1976) and opossums (Findlay, Elfont & Epstein,

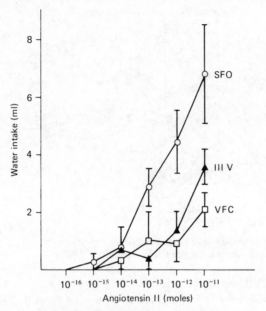

Fig. 4.12. Effect of angiotensin II on total water intake by the rat when injected into either the subfornical organ (SFO), the adjacent ventral fornical commissure (VFC), or adjacent dorsal third ventricle (IIIV). (From Mangiapane & Simpson, 1980.)

1980) with SFO lesions still drank after administration of angiotensin directly into the ventricles, which suggests that some area other than the SFO could also be involved. In the rat low doses of angiotensin stimulate drinking when injected into another circumventricular organ, the organum vasculosum of the lamina terminalis (OVLT) in the anteroventral region of the third ventricle (AV3V) (see figures 4.11 and 7.1). Lesions here also blocked drinking in response to angiotensin as well as to other thirst stimuli (Buggy & Johnson, 1977a), but these lesions were large and may have included damage to fibres from the SFO. This damage to SFO fibres could be responsible for the reduction in angiotensin-induced drinking since it was found that cutting the efferent fibres from the SFO to the OVLT or AV3V impaired drinking induced by angiotensin (Eng, Miselis & Salanga, 1980; Lind & Johnson, 1980). Studies in a larger animal such as the dog, in which lesions can be placed accurately and damage to surrounding tissue can be restricted, indicate that ablation of the OVLT alone did not affect drinking. However, if lesions extended beyond the OVLT to the nucleus medianus, drinking in response to angiotensin was reduced. Thus if there are receptors for angiotensin as a dipsogen in the region of the anterior third ventricle, they may be in an area other than the OVLT (Thrasher, Simpson & Ramsay, 1980).

During the controversy over whether the SFO or the OVLT might be the sole receptive site or the more important site for angiotensin dipsogenesis, Mogenson and his colleagues have continued to pursue the possibility that the preoptic area might contain angiotensin receptors. They found that bilateral injections of 0.5 ng (0.5×10^{-12} moles) of angiotensin through cannulae which were placed at such an angle that they would not pass through the ventricles stimulated drinking. Radioactive substances injected through the same cannula did not enter the ventricles. They propose that both the preoptic area and the circumventricular organs contain receptors for angiotensin dipsogenesis which respond in different ways since lesions in the lateral hypothalamus decreased drinking after injections in the preoptic area, but had little effect on drinking following ventricular injections (Swanson, Kucharczyk & Mogenson, 1978; Richardson & Mogenson, 1981).

Fitzsimons & Kucharczyk (1978) found that both the preoptic area and subfornical organ are very sensitive dipsogenic sites for angiotensin-induced drinking in the dog. The behaviour following stimulation of the preoptic area was not simply due to spread of the injectate to the ventricles because the magnitude of the response was greater and the latency to drink shorter with injections into the preoptic area than with ventricular injections. These findings also led to the conclusion that there is angiotensin-sensitive tissue in both the subfornical organ and the preoptic area.

A problem with the preoptic area as a receptive site is still that of the blood–brain barrier. To overcome this problem it has been suggested that the preoptic area might be responsive to angiotensin generated within the brain (Kucharczyk, Assaf & Mogenson, 1976). However, the current status of the proposal for a brain renin–angiotensin system is controversial (Hirose, Yokosawa & Inagami, 1978; Ramsay, 1979). The renin-like activity found in brain extracts does not occur at physiological pH (Day & Reid, 1976) and intracerebroventricular administration of naturally occurring renin substrate is relatively ineffective in stimulating drinking (Fitzsimons & Kucharczyk, 1978; Simpson, Reid, Ramsay & Kipen, 1978).

Part of the problem in interpreting the evidence for various receptor sites for angiotensin is that the interconnections of the several proposed receptive areas are not understood. It may be that when the functional anatomy is clearer the main controversies will resolve themselves. Such anatomical studies are now beginning to be made (Mogenson &

Kucharczyk, 1978). Swanson, Kucharczyk & Mogenson (1978) have shown that an efferent pathway from the angiotensin-sensitive sites in the preoptic area follows the medial forebrain bundle to the midbrain. Miselis, Shapiro & Hand (1979) found that the subfornical organ sends efferent connections to both the OVLT and the medial preoptic area. It may be that the three areas act as an integrated system for angiotensin-induced drinking.

Role of angiotensin in normal drinking

Another important issue concerning angiotensin is what role it plays in normal drinking. Is the drinking observed after angiotensin due to non-physiological manipulation of normal behaviour or does it indicate an important role for angiotensin in thirst?

We have already seen that injections of angiotensin produce highly motivated drinking in a wide variety of species. As yet, however, no physiological role for angiotensin in normal thirst has been proved (B. J. Rolls & Wood, 1977a; Stricker, 1978; Lee, Thrasher & Ramsay, 1981). One of the most common techniques to show that a system is involved in the control of a particular behaviour is to remove an essential component of the system and observe the behavioural changes. Thus, if we use water deprivation as our model for normal thirst, knowing that reduced extracellular volume contributes to the re-hydration, we might expect that removal or blockade of the angiotensin system would reduce drinking by an appropriate amount (i.e. 20–26 % at most, see table 4.4).

Since the level of circulating angiotensin is supposed to be an important influence on drinking, an obvious experiment was to remove the kidneys, the source of circulating angiotensin. To exclude the possibility that either the anuria or the surgery influenced intake, another group of rats had the ureters ligated. This was a particularly useful control because ureteric ligation is reported to elevate angiotensin levels. It can be seen in figure 4.13 that bilateral nephrectomy or ureteric ligation after overnight water deprivation did not differ in their effect on rehydration in spite of differences in circulating angiotensin levels. Further, the rehydration achieved after 3 h by these animals was similar to that of unoperated control rats. Thus water-deprivation-induced thirst was not affected by the removal of the renal renin–angiotensin system (B. J. Rolls & Wood, 1977a).

The physiological actions of angiotensin can be blocked by administering substances which occupy the receptors normally occupied by angiotensin. The administration of one such competitive blocker, saralasin acetate (P-113), also led to the conclusion that angiotensin is

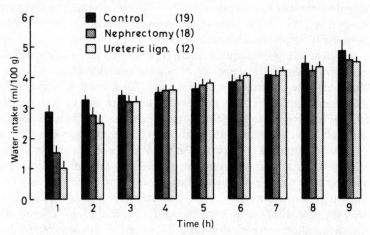

Fig. 4.13. The mean cumulative hourly water intake (ml/100 g initial body weight) after 21 h water deprivation by control rats which had not undergone surgery, by rats which were bilaterally nephrectomized, and by rats with both ureters ligated. (Numbers of animals in parentheses.) (From B. J. Rolls & Wood, 1977a.)

not essential for normal drinking following water deprivation. Injections or infusions of saralasin acetate into the cerebral ventricles of the dog (Ramsay & Reid, 1975), sheep (Abraham, Denton & Weisinger, 1976) or rat (Lee, Thrasher & Ramsay, 1981) did not affect water-deprivation-induced drinking.

There has been one report claiming a physiological role for angiotensin in deprivation-induced drinking (Malvin, Mouw & Vander, 1977), based on the finding that intraventricular infusion of saralasin acetate reduced such drinking; however, the infusions appear to have depressed the rats in a non-specific manner, perhaps through stress. This is evident in the data, in that when drinking was affected it was totally suppressed for varying lengths of time. Blocking the angiotensin system should not reduce deprivation-induced drinking by more than about 20–26 % since most drinking following deprivation is due to cellular dehydration (see pages 67–75).

Another approach has been to lesion the subfornical organ, and although this abolished drinking in response to intravenous angiotensin and delayed and sometimes reduced drinking in response to mildly hypertonic saline (Hosutt, Rowland & Stricker, 1981; Reed *et al.*, 1980), drinking following water deprivation was not affected (Kucharczyk, Assaf & Mogenson, 1976).

Thus the bulk of the evidence indicates that angiotensin is **not an** essential mediator of deprivation-induced water intake in the species studied so far in this context. That angiotensin is not essential does not necessarily imply that angiotensin does not normally play a role in water-deprivation-induced drinking. It may be that in drinking, as in other physiological systems, there is a redundancy of mechanisms so that when one is removed others take over (B. J. Rolls & Wood, 1977a). After deprivation the fluid changes are complex and various receptors may be involved. It may be that normally these receptors work together, but that if one type is removed the remaining receptors can take over the lost function. As it is essential to maintain fluid balance, redundancy would be an advantage. A recent experiment (Hoffman *et al.*, 1978) supports the redundancy hypothesis. It was found that blocking either angiotensin receptors or cholinergic receptors (which they suggested may mediate drinking induced by cellular dehydration), had no effect on deprivation-induced drinking. However, combined blockade of both receptors reduced deprivation-induced drinking by 70%. They suggest that the two receptors are independently capable of maintaining thirst.

If the removal experiments do not provide conclusive evidence that angiotensin is involved in deprivation-induced thirst, a remaining approach is to manipulate angiotensin levels and to measure the effects on behaviour. Angiotensin has been infused both centrally and peripherally in rats and dogs which were in fluid balance and has been found to stimulate significant drinking at levels which could occur physiologically (Trippodo, McCaa & Guyton, 1976; Hsaio, Epstein & Camardo, 1977; Fitzsimons, Kucharczyk & Richards, 1978). Thus there is no doubt that angiotensin is an extremely potent dipsogen. It should be pointed out that if injections or infusions are used to assess the physiological significance of substances such as angiotensin the route of administration should be chosen to mimic the release of the naturally occurring compound. For angiotensin intravenous infusions are more likely to represent the normal physiology where renal renin is released and angiotensin is formed in the peripheral plasma. It has been suggested that the central administration of angiotensin has little physiological relevance. Angiotensin II is present in the cereorospinal fluid only in very low concentrations (which are lower than those in plasma) and it does not normally enter the cereorospinal fluid from plasma (Ramsay & Reid, 1975; Findlay, Elfont & Epstein, 1980).

The elevation in plasma angiotensin after water deprivation is similar to that which results from dipsogenic doses of intravenous angiotensin (Abdelaal, Mercer & Mogenson, 1976; Mann, Johnson & Ganten, 1980). We have recently found that renin activity is also elevated by water deprivation in man and that it decreases towards normal levels during rehydration (B. J. Rolls, Wood, Rolls *et al.*, 1980). Whether angiotensin is involved in normal drinking or not, it seems possible that the renin–angiotensin system could provide an emergency mechanism for protecting plasma volume in some pathological conditions (see Chapter 9).

Angiotensin and sodium appetite

We have seen that depletions of the extracellular fluid compartment, in particular of the vascular compartment, lead to highly motivated water consumption in a variety of species. However, water on its own is not entirely appropriate for the replenishment of extracellular fluid since depletions of sodium must also be made up. It has been known since 1936 (Richter) that a specific appetite for sodium can develop in animals where there are disorders of sodium metabolism (e.g. after adrenalectomy). It has also been observed that, following extracellular fluid depletions, rats develop a sodium appetite if no food is available during rehydration (in a normal situation food would provide adequate sodium to replenish losses). The renin–angiotensin system mediates some of the water intake which follows extracellular depletion and there is now convincing evidence that the renin–angiotensin system is also important in sodium appetite. Long-term infusions of angiotensin II into the third cerebral ventricle led to an increased intake of normally aversive hypertonic saline solutions (see figure 4.14) and to a positive sodium balance (Avrith & Fitzsimons, 1980). Not only did rats drink large volumes of hypertonic salt solutions while they were given angiotensin infusions, but some animals continued to drink sodium chloride after the termination of the infusions. This could imply that prolonged elevations of angiotensin may alter the mechanisms for salt perception, enhancing the hedonic value of strong solutions and making them more acceptable (Bryant, Epstein, Fitzsimons & Fluharty, 1980).

The phenomenon of sodium appetite is currently under investigation in a number of laboratories. It is still not clear whether sodium appetite is due to the salt deficiency, the decreased availability of sodium in the brain (Stricker, 1981) or the endocrine consequences of sodium deficiency. Some of the outstanding problems for those who favour an endocrine explanation are whether levels of angiotensin in the brain can entirely account for sodium appetite or whether circulating angiotensin also plays a role. The mineralocorticoids, for example aldosterone, have also been implicated in sodium appetite and the role of these hormones and the way in which they might interact with

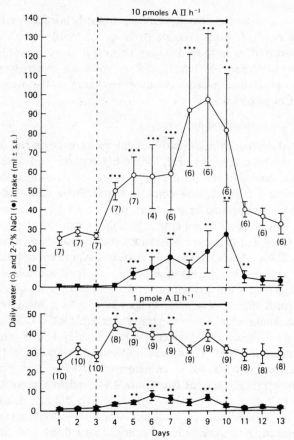

Fig. 4.14. Sodium appetite produced by the infusion (days 3–10) of angiotensin II (AII) into the third cerebral ventricle. (Numbers of rats in parentheses; single asterisk, $P<0.05$; double asterisk, $P<0.01$; triple asterisk, $P<0.001$, compared with the three pre-infusion base-line days.) (From Avrith & Fitzsimons, 1980.)

angiotensin need to be clarified. Little is known about the brain areas involved in sodium appetite. Also, although salt appetite is clearly a useful adaptive mechanism for emergencies, little is known about the importance of such an appetite in normal circumstances. The books *The Physiology of Thirst and Sodium Appetite* (Fitzsimons, 1979) and *Biological and Behavioral Aspects of Salt Intake* (ed. Kare, Fregly & Bernard, 1980) provide comprehensive recent reviews of this subject.

How angiotensin might act to stimulate drinking

The way in which angiotensin acts to stimulate drinking is still not clear and this is an important area for future research. Since physiologists are always eager to understand bodily mechanisms it is perhaps not surprising that there has been considerable speculation as to the mode of action of this powerful dipsogen. One of the earliest suggestions was that angiotensin might make stretch or pressure receptors in the vasculature more sensitive to decreased plasma volume. This information would then be sent to the central nervous system and drinking would follow (Fitzsimons, 1969). Since drinking is induced more easily by injecting angiotensin directly into the brain than by infusing it intravenously, the possibility of systemic mechanisms for angiotensin has had little experimental attention.

Most of the speculation about the mode of action of angiotensin has stemmed from the observed sensitivity of the brain to minute quantities of angiotensin. One of the most obvious possibilities is that there are specific thirst receptors which respond to changes in angiotensin concentration. This is supported by findings that modifications of the molecular structure of angiotensin when delivered intracranially affect its dipsogenic potency. Also, various inactive compounds which compete for the receptor block drinking in response to angiotensin (see Fitzsimons, 1979). In the search for the location of the angiotensin receptors most attention has been focussed on the subfornical organ. The SFO appears to be organized with a superficial layer of neurons which would be accessible to substances such as angiotensin II in the cerebrospinal fluid or blood. These superficial neurons project to deeper cholinoceptive neurons which pass out of the SFO in the fornix. Electrophysiological studies indicate that SFO cells show specific sensitivity to small quantities of angiotensin, which have been demonstrated *in vitro* to be within the physiological range. It was shown *in vivo* (figure 4.15) that the response of the SFO neurons was

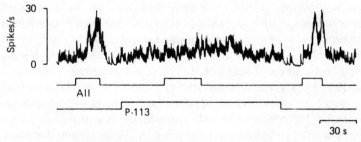

Fig. 4.15. The effect of angiotensin II (AII) on a cell (or cells) in the cat subfornical organ that responded only to angiotensin, and the effect on this response of the ejection of the competitive angiotensin inhibitor saralasin acetate (P-113). (From Phillips & Felix, 1976.)

related to the dose applied and that the response could be blocked by saralasin acetate (P-113), a competitive angiotensin antagonist (Felix & Akert, 1974; Phillips & Felix, 1976; Felix & Schlegel, 1978). Although it has been suggested that the SFO might also contain osmosensitive neurons, this was not supported by *in vitro* studies which showed that changes in sodium concentration did not affect neuronal responsiveness. Studies of the SFO have revealed the interesting possibility that the SFO is not just a receptive site for angiotensin; it could be a neuro-secretory organ involved in the control of fluid balance. The morphology of the cells is suggestive of neurosecretion; extracts of SFO affect the fluid balance of rats; and dehydration and haemorrhage affect the morphology of the SFO (see Buranarugsa & Hubbard, 1979). Thus, although it is not proven that the SFO contains receptors for angiotensin-induced drinking (proof requires that angiotensin binding and neuronal responsiveness are shown to occur together and that these responses are associated with drinking), the evidence indicates that this is a promising hypothesis to pursue.

Because angiotensin is a powerful vasoconstrictor and because two of the most sensitive brain areas for angiotensin dipsogenesis, namely the SFO and the OVLT, are highly vascularized, it has been suggested that drinking is due to local vasoconstriction in these areas caused by increased levels of circulating angiotensin (see Fitzsimons, 1979). According to this view the SFO (and perhaps the OVLT) would act as receptors for extracellular fluid volume. Constriction would be detected by stretch receptors and interpreted as decreased filling, and thirst would result. In support of this view, vasoplegics (substances which cause paralysis of blood vessels), such as prostaglandin E, block drinking induced by angiotensin when injected into the ventricular organs (Kenney & Epstein, 1978). Also, a variety of substances which have little in common apart from an action on blood vessels have effects on drinking. The theory is challenged, however, by the findings that not all vasoconstrictors cause drinking and that not all vasodilators oppose the effects of angiotensin on thirst (Phillips & Hoffman, 1977).

The possibility has also been mooted that angiotensin is not simply a blood-borne hormone acting on the central nervous system, but that it might be a neural transmitter substance. Evidence which could support this view comes from immunohistofluorescence studies which indicate angiotensin II-like activity in widely distributed neurons. This neuro-transmitter theory is made more plausible by the suggestion that there is a complete system within the brain itself for the formation of renin and angiotensin. As has already been pointed out, however, the existence of this intrinsic renin–angiotensin system in the brain is still

Table 4.2. *Theories on the mechanism of angiotensin-induced thirst*

Theory	Further discussion
angiotensin acts by sensitizing peripheral vascular stretch or pressure receptors to hypovolaemia	Fitzsimons, 1969
angiotensin receptors are located in the brain	Phillips, 1978; Buranarugsa & Hubbard, 1979; Felix & Schlegel, 1978
'local ischaemia hypothesis': angiotensin acts by causing local vasoconstriction in the SFO and OVLT	Fitzsimons, 1979
angiotensin is a neurotransmitter	Fitzsimons, 1978
angiotensin and sodium interact on the thirst receptor	Andersson, 1978

controversial as is the suggestion that angiotensin could be a neuro-transmitter (see Fitzsimons, 1978 and 1979 for further discussion). Even if angiotensin were shown to be a neurotransmitter it is possible that this function could be separate from that of angiotensin in the systemic circulation, where it appears to act as a hormone for the regulation of blood volume (see figure 2.7).

Another hypothesis is that angiotensin and sodium are both activators of an enzyme involved in excitation of the thirst receptor (see Andersson, 1978, for review). The basis of this hypothesis is that the combined administration of angiotensin and sodium ions results in a 'drastic augmentation' of the dipsogenic effect of either stimulus given alone. This hypothesis relies on the thirst receptors being sodium receptors and, as discussed earlier, this is unlikely to be the case. Also in the *in vitro* electrophysiological studies of the SFO, one of the most likely areas for angiotensin receptors, it was found that 20% changes in osmotic pressure with sodium chloride had no effect on the responsiveness of cells to angiotensin (Buranarugsa & Hubbard, 1979). Although there is no proof that angiotensin and sodium interact on the thirst receptor, there is a recent report that sodium balance affects the dipsogenic response to angiotensin. Sodium-depleted rats drink less than normal after a given dose of angiotensin (Mann *et al.*, 1979). The influence of sodium deprivation is not, however, very general in that the effects of central angiotensin on blood pressure and blood flow are the same in normal and in sodium-depleted dogs (Brosnihan, Berti & Ferrario, 1979).

Andersson (1978) suggests that all theories about the mode of action of any thirst receptor must remain pure speculation until we know more about the way in which such receptors are normally influenced.

Important questions are whether the blood or the cerebrospinal fluid composition is the more important influence on thirst and how the blood–brain barrier affects the composition of these fluids (for further discussion see Thrasher, Brown, Keil & Ramsay, 1980; Thrasher, Jones, Keil, Brown & Ramsay, 1980).

This section is aimed primarily at introducing the range of ideas (summarized in table 4.2) that scientists are currently exploring to try to understand how minute quantities of angiotensin can produce such specific and highly motivated drinking behaviour. The various theories discussed are not necessarily mutually exclusive. It is this type of theorizing which sometimes leads scientists in related areas to develop techniques which make possible the direct testing of a particular hypothesis.

Cardiac receptors for thirst

It has been supposed for some time that not all of the drinking following hypovolaemia can be accounted for by increases in the level of circulating angiotensin. Removal of both kidneys, which eliminates changes in the level of circulating angiotensin, has little effect on drinking following administration of polyethylene glycol (see Stricker, 1973) or water deprivation (B. J. Rolls & Wood, 1977a, see figure 4.13). Local changes in blood volume and pressure in and around the heart can affect the release of antidiuretic hormone and it has been thought likely that receptors in this region could also be involved in hypovolaemic drinking. These receptors may be manipulated by inflating small chronically implanted balloons in the vasculature around the heart. This technique has now been applied to the study of thirst. Following up the finding of Ramsay, Rolls & Wood (1975) that partial constriction of the thoracic inferior vena cava stimulates drinking in dogs, Fitzsimons & Moore-Gillon (1980b) inflated a balloon (see balloon A in figure 4.16) in the abdominal inferior vena cava. This balloon caused a decrease in the blood pressure and volume in and around the heart. Such changes should be interpreted by putative cardiac receptors as a decrease in blood volume, and increased fluid intake as well as antidiuresis should occur in order to replace and conserve plasma volume. These responses were observed in the dogs tested and the drinking during the first hour correlated with the fall in peripheral arterial pressure (see figure 4.17). More critically, there was also a highly significant positive correlation between the amount drunk and the maximum fall in central venous pressure. If the balloon remained inflated for 1–3 days there was a sustained increase in water intake and an indication of a delayed sodium appetite. Blood volume

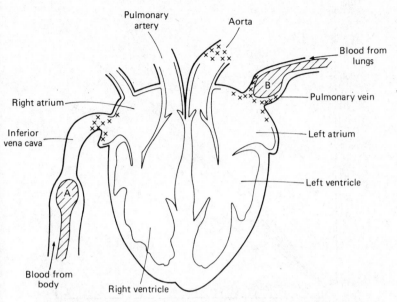

Fig. 4.16. Schematic diagram of the heart with crosses indicating possible locations of some of the volume and pressure receptors. Inflation of balloon A in the inferior vena cava reduces blood volume and pressure in and around the heart and leads to drinking. Inflation of balloon B at the entry of the pulmonary vein to the left atrium increases the pressure in that region and inhibits drinking.

and pressure were not only reduced in and around the heart but also around the kidneys. Fitzsimons & Moore-Gillon (1980b) found that plasma renin levels were elevated, so the drinking could be due to stimulation of either the cardiac receptors, the renin–angiotensin system, or both. Ramsay, Rolls & Wood (1975) found that the persistently elevated water intake seen after thoracic caval constriction was reduced by the competitive angiotensin inhibitor saralasin acetate. Similarly, saralasin acetate reduced the drinking following caval balloon inflation so at least part of the drinking following these procedures was angiotensin dependent. There was, however, some residual drinking that was not angiotensin dependent and presumably resulted from activation of cardiac receptors.

Thus decreased activation of cardiac receptors leads to increased drinking. It seems possible, then, that a stimulus interpreted as increased cardiac filling might decrease fluid intake. To test this Fitzsimons & Moore-Gillon (1980a) inserted a balloon (see balloon B in figure 4.16) so that it was located at the entry of the pulmonary vein to the left

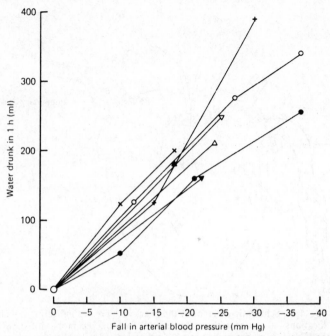

Fig. 4.17. Water intake in response to graded obstruction of the inferior vena cava. Different symbols represent different dogs. The control (large open circle) is the mean of eight dogs. (From Fitzsimons & Moore-Gillon, 1979.)

atrium. This is a site where vascular receptors are located so that distension of a balloon here would stretch the receptors and this stretching would be interpreted as an increase in blood volume and pressure. Thus the appropriate behavioural response in this situation would be a decrease in fluid intake. It was found that dogs made thirsty by injection of the beta-adrenergic agonist isoproterenol (see Chapter 8) drank significantly less when the balloon was inflated (191.1±37.4 ml) than when it was not (266.3±41.6 ml). Interestingly, if the balloon was left inflated for 24 h the *ad libitum* water intake of the dogs was reduced.

Thus it appears that there are receptors in and around the heart which could respond to changes in blood volume or pressure and which could be involved in the control of drinking. It is not, however, clear precisely where the receptors are located. Fitzsimons & Moore-Gillon favour loci in the low-pressure (venous) circulation around the heart since the compliance of these vessels is high, making them responsive to changes in blood volume. Another group has also suggested that the receptors are in and around the left side of the heart, in particular in the left atrium (Zimmerman, Blaine & Stricker, 1981). It appears that the receptors involved in the control of ADH release are quite widely distributed, and this may also be the case for thirst receptors. More work is required to characterize the receptors.

The vagus nerve (which contains afferent fibres for atrial receptors, aortic baroreceptors, aortic bodies and other parts of the circulation) appear to provide a link between the cardiac receptors and the central nervous system (Sobocińska, 1969). The afferent impulses in this neural path appear to exert an inhibitory effect on thirst. Thus, blocking the vagosympathetic trunk enhanced the drinking in response to balloon inflation in the inferior vena cava (balloon A, figure 4.16) (Fitzsimons & Moore-Gillon, 1980b) and abolished the reduction in drinking caused by inflation in the pulmonary vein (balloon B, figure 4.16) (Fitzsimons & Moore-Gillon, 1980a).

Summary – extracellular fluid depletion as a thirst stimulus

A reduction in plasma volume (hypovolaemia) leads to fluid conservation by the kidneys and to increased fluid intake. Receptors which could monitor changes in plasma volume appear to be located in and around the kidneys and the heart. The role of the kidney in extracellular thirst has had much experimental attention and the major findings are

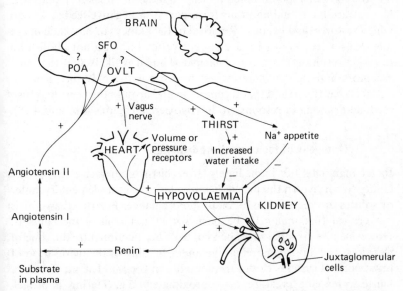

Fig. 4.18. A summary of the mechanisms involved in extracellular thirst. POA, preoptic area; SFO, subfornical organ; OVLT, organum vasculosum of the lamina terminalis.

summarized in figure 4.18. When hypovolaemia is sensed in the juxta-glomerular apparatus in the kidney, the enzyme renin is released. Renin acts on a substrate in the plasma to form angiotensin I which is converted to the physiologically active form, angiotensin II. Angiotensin II is an extremely potent dipsogen when infused intravenously or when injected directly into the brain. The effective doses are thought to be within the physiological range (i.e. the levels found when drinking would normally occur, such as after water deprivation). The importance of angiotensin in normal drinking which occurs spontaneously or after deprivation is not yet clear, but it is likely that angiotensin is important as a dipsogen in some clinical conditions (see Chapter 9). The receptors for angiotensin-induced thirst may be located in several brain areas. For example, several of the circumventricular organs (the subfornical organ and the organum vasculosum of the lamina terminalis) which lie outside the blood-brain barrier are highly responsive to angiotensin, and lesions in the subfornical organ abolish drinking in response to intravenous angiotensin. There may also be receptors inside the brain, for example in the preoptic area, although the fact that angiotensin does not cross the blood–brain barrier makes it difficult to see how angiotensin would normally reach these sites. Angiotensin produces highly motivated drinking behaviour in a wide range of species. The ingestion of water and the sodium appetite which follow angiotensin infusions replace the lost extracellular fluid and thus form part of a negative-feedback loop which restores fluid balance. Removal of the kidneys or blockade of the angiotensin receptors with a competitive inhibitor do not abolish all drinking resulting from hypovolaemia. Thus it is likely that the renal receptors and the angiotensin system are not the only mediators of extracellular thirst, and that receptors in and around the heart may also mediate drinking in response to hypovolaemia (figures 4.16 and 4.18).

Ontogeny of the controls of drinking

In all mammals life immediately after birth is characterized by the ability to survive on the mother's milk alone. No supplementary water or solutes are required to maintain fluid balance. For this reason it is not critical that mammals be born with thirst control mechanisms. However, it is essential that such controls be functional by the time of weaning. The most comprehensive understanding of the maturation of thirst control systems comes from studies in the rat. The rat is born a hairless, helpless creature of approximately 5 g. During a 3-week lactation period it grows rapidly and becomes ready to regulate its own food and water intake by the time of weaning. To study the development of regulatory drinking Wirth & Epstein (1976) devised a water fountain that enables new born pups which would not normally drink water to lap from the spout in response to thirst challenges. They found precocious development of thirst systems, with the infant rats responding to all of the stimuli tested within the first week of life. The pups drank water in response to cellular dehydration (produced by subcutaneous hypertonic sodium chloride) at 2 days of age, to hypovolaemia (produced by subcutaneous polyethylene glycol) at 4 days of age (Wirth & Epstein, 1976), and to angiotensin (injected into the ventricles) at 4 days of age (Epstein, 1978; Misantone, Ellis & Epstein, 1980). Rat pups are capable of showing appetitive behaviour at a surprisingly early age. For example, they respond to cellular dehydration by lapping water off the floor of the cage at 3 days of age (Bruno & Hall, 1980). Thus, the neurophysiological controls for drinking are functional well before they are essential for survival. Capacities for mastery of the habitat including improved locomotion, use of distance senses, and some competence for thermoregulation appear at the end of the second post-natal week. It is at this time that the pups become capable of obtaining fluids from containers such as bowls.

There is some evidence that the early drinking experiences of rats can influence later drinking behaviour. For example, rats which before adulthood never had access to drinking water, but which had to meet fluid requirements by eating lettuce, as adults drank little in response to an injection of hypertonic saline (Milgram, 1979). Also, pups of mothers deprived of sodium during pregnancy and early lactation showed a 2% augmentation of water intake as adults (Mouw, Vander & Wagner, 1978). It is possible that early experience with salt and water can cause permanent changes in the neural controls of drinking.

Although rat pups have the capacity to respond to fluid deficits by drinking water at an early age, this does not imply that suckling of milk (the source of both food and water) is controlled by drinking rather than by feeding mechanisms. To investigate this important issue, Bruno (see Blass, Hall & Teicher, 1979) dehydrated rat pups and assessed suckling behaviour. He found that before 2 weeks of age suckling of milk was not influenced by dehydration. After 2 weeks of age dehydration reduced suckling, just as it is known to decrease feeding in adults. Therefore it appears that suckling of milk is not controlled by thirst mechanisms, but is more likely to be a feeding behaviour.

It is not clear what the relevance of the early sequential development of thirst mechanisms in the rat is to other species. It would be particularly important to have more understanding of the development

Table 4.3. *Plasma changes after water deprivation*

	Rat (n = 15)	Dog (n = 8)	Monkey (n = 5)	Man (n = 5)
Osmolality (mosmoles/kg H_2O)				
non-deprived	299.6±0.5	298.5±0.7	297.8±2.4	282.4±2.2
deprived	306.1±0.6[a]	310.3±0.9[a]	311.0±4.5[a]	290.8±1.8[a]
Sodium (mequiv./l)				
non-deprived	138.9±0.5	147.3±0.7	143.0±1.7	140.4±0.7
deprived	139.9±0.5	153.7±0.7[a]	149.2±2.0[a]	143.3±0.6[a]
Haematocrit (%)				
non-deprived	42.2±0.7	42.4±2.0	36.6±1.9	47.2±1.8
deprived	46.2±0.6[a]	47.3±2.1[a]	35.8±1.1	48.2±2.3
Plasma protein (g%)				
non-deprived	7.5±0.2	5.5±0.1	7.1±0.2	7.3±0.2
deprived	7.8±0.1[a]	6.1±0.1[a]	7.4±0.2	7.7±0.2[a]

The periods of water deprivation (with food available) were 21 h for the rat and 24 h for the dog, monkey and man. Osmolality and sodium provide indices of cellular dehydration. Haematocrit and plasma protein indicate changes in plasma volume. Significant differences between predeprivation and postdeprivation values are indicated; †[a], $P < 0.001$. (From Rolls, Wood & Rolls, 1980.)

of the regulatory capacities of newborn humans. This understanding could be important clinically since in recent times the high incidence of early formula and solid-food diets has meant that infants may be taking in more concentrated protein and salt than the kidneys can manage. A study of plasma osmolality (Davies, 1973) indicated that the osmolality of breast-fed babies aged 1–3 months was 284 mosmoles/l, whereas that of babies being fed both milk formula and solid foods was 297 mosmoles/l. It was not reported whether these hyperosmotic infants were offered supplementary water and, if so, whether they would drink it. The situation could be complicated by the fact that water is aversive to the newborn infant, and from the first day of life taste appears to be a crucial determinant of intake (Desor, Maller & Andrews, 1975). It is important to understand more about the early controls of drinking because hypertonicity in infants can be dangerous in illness leading to vomiting, diarrhoea or decreased feeding. Extreme dehydration can lead to brain damage and sometimes death (see Chapter 9). Also, it has been suggested that excessive sodium chloride consumption in infancy can lead to an acquired taste for sodium chloride. High sodium chloride intake can lead to expanded extracellular fluid volume and may relate to a higher prevalence of hypertension (Freis, 1976).

The initiation of drinking following water deprivation

It is important to study drinking following water deprivation in detail because deprivation is a relatively natural way to produce thirst and could reveal the physiological changes which might act to stimulate drinking in animals in normal non-laboratory situations. In the standard experiment that we will describe, water but not food was withdrawn for 21–24 h. The work indicates that such thirst is initiated by depletion of both the cellular and extracellular fluid compartments.

The effect of water deprivation on the cellular and extracellular fluid

The effect of the 21–24 h water deprivation on body fluid balance is shown for the four different species we have investigated (table 4.3) (data from Ramsay, Rolls & Wood, 1977a, b; Wood, Maddison, Rolls, Rolls & Gibbs, 1980; B. J. Rolls, Wood, Rolls *et al.*, 1980). In the rat, dog, monkey and man it is clear that water deprivation depletes both the cellular fluid compartment (indicated by the elevation of plasma osmolality or sodium concentration) and the extracellular fluid compartment (assessed by the elevated haematocrit or raised plasma protein levels). Thus both cellular and extracellular stimuli of thirst could contribute to the drinking following water deprivation.

Selective removal of the cellular or extracellular deficits

Given that the cellular and extracellular fluid compartments are depleted by water deprivation, the next question is to what extent these depletions actually contribute to the drinking following water deprivation. This has been investigated in laboratory animals by selective replacement of the deficits produced by water deprivation, using infusions of water to replace the cellular deficit and infusions of isotonic solution to re-expand the extracellular fluid volume (see figure 4.19). After it had been confirmed by measurements of plasma variables that the selective replacement produced by the preload had been successful, the residual drinking was measured.

Preloads in the deprived rat

A summary of the results of the preload experiments on the three species tested is given in table 4.4. In the rat the selective replacement was achieved by administering preloads of water or a balanced isotonic salt solution by three different routes, i.e. intravenously, intragastrically, or orally. The results indicate that the route of administration of the

Table 4.4. *The effect of preloads on drinking following water deprivation*

	% reduction in water intake[a]		
	Rat ($n = 15$)	Dog ($n = 8$)	Monkey ($n = 3$)
Removal of cellular thirst stimulus	64–69	72[b]	85
Removal of extracellular thirst stimulus	20–26	27	5

[a]Water intake is expressed as a percentage of the intake in control conditions where the deficits are not restored. (From B. J. Rolls, Wood & Rolls, 1980.)
[b]Only the central osmoreceptors were rehydrated.

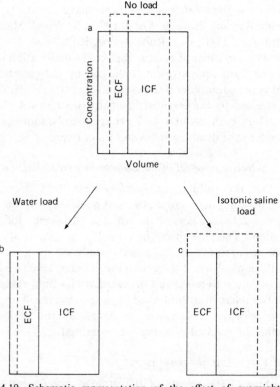

Fig. 4.19. Schematic representation of the effect of overnight water deprivation on the cellular (ICF) and extracellular (ECF) fluid compartments. The normal isotonic state is represented by the solid lines. The dotted lines indicate the direction of changes after (a) deprivation and after the equilibration of preloads of (b) water (which leaves the ECF depleted) or (c) isotonic saline (which leaves the ICF depleted).

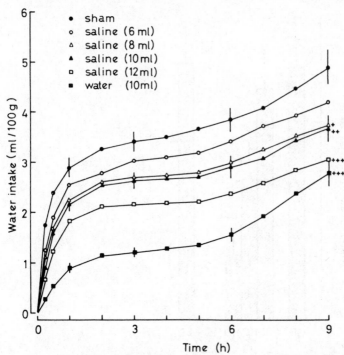

Fig. 4.20. The mean cumulative water intake of 19 water-deprived rats, used as their own controls, after intragastric preloads of 6, 8, 10, or 12 ml balanced salt, 10 ml deionized water, or a sham load. ($+$, $P<0.05$; $++$, $P<0.01$; $+++$, $P<0.001$.) (From Ramsay, Rolls & Wood, 1977b.)

preload had little effect on intake. A preload of 10 ml water which brought plasma osmolality down to predeprivation levels but had little effect on plasma volume reduced the total intake of water in 1 h by 64–69% (table 4.4). If plasma volume was restored with an isotonic salt solution which had little effect on osmolality, drinking was reduced by 20–26% after 1 h. This reduction in drinking was a graded effect, correlating with the volume of the preload (figure 4.20). Thus, in the rat after water deprivation, changes in both the cellular and extracellular fluid compartments contribute to drinking.

Preloads in the deprived dog

Our technique for studying deprivation-induced drinking in the dog was somewhat different from that in the rat in that the larger size of the animal allowed the preparation of carotid loops (Ramsay, Rolls & Wood, 1977a). Since the carotid arteries provide the main blood supply to the brain, we were able to manipulate the osmolality of the blood

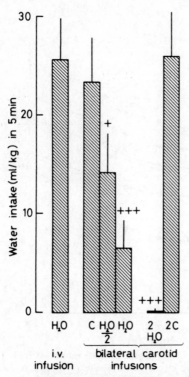

Fig. 4.21. The dose-dependent reduction in drinking by eight water-deprived dogs infused via the carotid arteries with water. H_2O, the infusion of water which restored central plasma osmolality to pre-deprivation levels; $H_2O/2$, infusion of water at half that rate; $2H_2O$, infusion of water at twice that rate. C, infusion of isotonic saline at the same rate as water. Bilateral carotid infusions of C or 2C did not influence drinking nor did the intravenous (i.v.) infusion of water. ($+$, $P<0.05$; $+++$, $P<0.001$.) (From Ramsay, Rolls & Wood, 1977a.)

perfusing the brain without changing significantly the concentration of the blood in the rest of the body. The tonicity of the blood perfusing the brain was monitored via a catheter in the jugular vein, the main exit route of blood from the brain. Peripheral plasma changes were monitored via a catheter in a leg vein.

After overnight deprivation of water, bilateral intracarotid infusions of water at a rate which brought jugular plasma osmolality down to predeprivation levels reduced water intake by 72% (table 4.4) compared to control infusions of isotonic saline. The effect of higher and lower rates of infusion showed the reduction in drinking to be dose related (figure 4.21). Since control infusions of isotonic saline at different rates did not affect water intake and the dogs would eat normally if offered food during water infusions, there was no indication that the reduction of water intake was due to discomfort. Analysis of peripheral venous blood samples showed that the carotid infusions did not significantly change systemic plasma osmolality. Also, infusing water at the same rate intravenously did not reduce intake, so it is concluded that the reduction of drinking was due to removal of the central stimulus. The

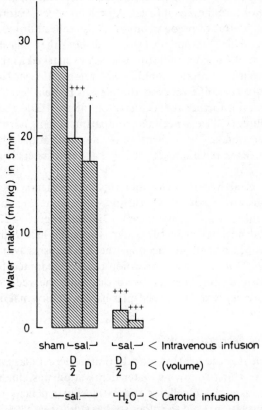

Fig. 4.22. The three bars on the left show the reduction of drinking in eight water-deprived dogs following the restoration of the extracellular fluid compartment to predeprivation levels with an intravenous infusion of isotonic saline (sal.). The two bars on the right show the virtual abolition of deprivation-induced drinking by the combined restoration of central plasma osmolality and the ECF volume to predeprivation levels. D, volume of saline equal to fluid deficit calculated from change in body weight. This volume of saline over-expanded the ECF whereas infusion of saline at half this rate (D/2) restored the ECF. H_2O, the bilateral carotid infusion of water which restored central plasma osmolality to predeprivation levels. (+, $P<0.05$; +++, $P<0.001$). (From Ramsay, Rolls & Wood, 1977a.)

72% reduction in water intake is particularly impressive when it is remembered that, because systemic cellular dehydration and hypovolaemia persisted, the dogs were still out of fluid balance when they terminated their drinking. Thus, central cellular dehydration appears to be the factor which accounts for the greater part of drinking after water deprivation in the dog.

To test whether the reduction in plasma volume seen after deprivation (table 4.3) contributes to the drinking following water deprivation in the dog, a preload of isotonic saline was infused. A preload which restored the extracellular fluid volume to predeprivation levels reduced drinking by 27% (table 4.4). This isotonic infusion did not alter plasma osmolality. This effect, although significant, was not as marked as that observed when the central osmotic stimulus was removed. Doubling the volume of the infused saline reduced the plasma protein level to below normal, but did not further reduce drinking (figure 4.22). Thus, restoration of the volume of the extracellular compartment to its normal level reduced water intake but there was no indication that hypervolaemia inhibited the remaining thirst (i.e. that due to a persisting cellular stimulus).

When both the central hypertonicity and the hypovolaemia were removed simultaneously by giving both carotid water infusions and intravenous saline, there was no significant drinking (figure 4.22). Half the dogs showed no interest in the water; the other half drank small amounts. It appears that, in the dog, the mechanisms which control drinking after 24 h of water deprivation can be accounted for in terms of the raised central plasma osmolality and the reduced extracellular fluid volume, with the raised plasma osmolality making the larger contribution.

Preloads in the deprived monkey

In both the rat and the dog the relative contribution of the cellular and extracellular deficits to thirst following water deprivation was similar (table 4.4). In the monkey, both cellular and extracellular deficits resulted from water deprivation, as in the other species studied (table 4.3). The contribution of these changes to thirst has been studied in monkeys (Wood, Rolls & Rolls, 1982) that were deprived of fluid for 24 h and then infused with different volumes of water in order to rehydrate the cellular fluid compartment to different extents before the animals were allowed to drink. The effect of the infusions on drinking was dose-related in a linear fashion (figure 4.23). Infusion of a volume of water which returned plasma sodium and osmolality to predeprivation levels reduced drinking by a mean value of 85% (table 4.4). Larger infusions reduced plasma

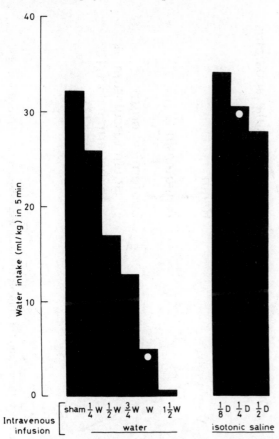

Fig. 4.23. The reduction in drinking in the water-deprived monkey (three animals) by the intravenous infusion of water (W) or isotonic saline (D) at different rates. White circles indicate treatments that restore measures of cellular hydration (water infusion) or extracellular hydration (isotonic saline infusion) to predeprivation values. (From B. J. Rolls, Wood & Rolls, 1980.)

concentration to below normal and virtually abolished drinking. Infusions of isotonic saline were also given to attenuate selectively the extracellular fluid deficit before allowing the animals to drink. A volume of saline which restored plasma volume to predeprivation levels reduced the mean water intake by only 5% (table 4.4) in the monkeys. A smaller volume of saline had no effect, and larger volumes which expanded plasma volume above normal without debilitating the animals produced a similarly small further reduction in intake (figure 4.23). Thus, in the monkey the cellular stimulus to thirst is even more important in relation to the extracellular stimulus than in the rat or dog.

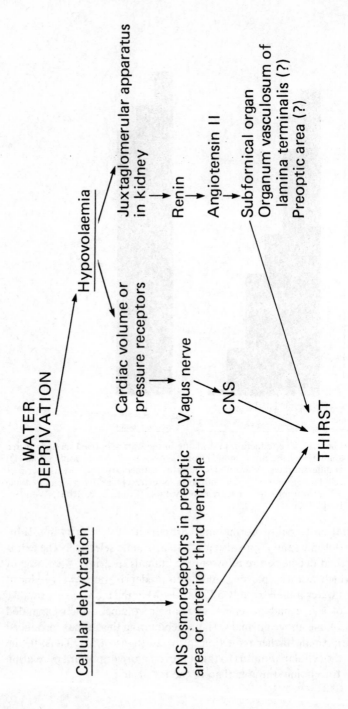

Fig. 4.24. A summary of the factors which may contribute to the initiation of drinking following water deprivation. (From B. J. Rolls & Rolls, 1981.)

In man the role of the observed cellular and extracellular fluid deficits in initiating drinking after water deprivation has not been definitively established by stimulus removal experiments such as those we have performed in the rat, dog and monkey. However, by infusing hypertonic sodium chloride solution Wolf (1950) calculated that humans drank when intracellular fluid volume was decreased by 1.2%. Since we found that the change in sodium (2.1%) after water deprivation was well above that threshold (see Table 4.3), it is likely that the cellular fluid deficit is an important component of deprivation-induced drinking in man.

Summary – factors which initiate deprivation-induced drinking

Following overnight fluid deprivation both the cellular and extracellular fluid compartments are significantly depleted. Preloads of water which replete the cells but have insignificant effects on the extracellular compartment reduce drinking by 64–72% in the rat and dog and by 85% in the monkey. Restoring the extracellular compartment to pre-deprivation levels reduces drinking by 20–27% in the rat and dog, but by only 5% in the monkey. Thus cellular dehydration is the most important factor, but at least in some species reductions in the extracellular fluid volume contribute to deprivation-induced drinking. A summary of the factors which may contribute to the initiation of drinking following deprivation is shown in figure 4.24.

5   The maintenance of drinking

So far we have considered factors which can initiate drinking. We now consider how drinking is maintained. Do animals drink in order to remove thirst stimuli? For example, do they drink in order to dilute their plasma or expand their extracellular fluid volume? Or do animals drink in order to obtain oropharyngeal sensations such as the taste of water?

The role of oropharyngeal, gastric and intestinal factors, and post-absorptive factors in the maintenance of drinking.

One way in which these factors can be separated is in the sham-drinking preparation, in which, for example, the animal may drink water and receive oropharyngeal sensation but other factors do not operate because the ingested water drains through an oesophageal or gastric fistula (see figure 3.4). It is found that such sham drinking is maintained, i.e. at least as much water is ingested when the fistula is open as when it is closed. In fact, the drinking is maintained very convincingly, in that animals usually sham drink very large quantities of water. This has been shown in the horse (Bernard, 1856), the dog (Bernard, 1856; Bellows, 1939; Towbin, 1949; Adolph, 1950) and more recently in the rat (Mook, 1963; Blass & Hall, 1976). An example of the type of sham drinking which can be demonstrated in the 18-h water-deprived rat with a gastric fistula open is shown in figure 5.1 (from experiments of B. J. Rolls, E. A. Rowe, J. G. Gibbs, E. T. Rolls & S. Maddison). In fact, animals drink to obtain oropharyngeal stimulation by water, even when no water reaches the intestine or when no water is absorbed. Thus, oropharyngeal factors such as the taste of water and perhaps swallowing are sufficient to maintain drinking, i.e. to provide the incentive or reward for drinking, in these species. It is not necessary for water to reach the intestine, or for it to be absorbed.

Although the importance of oropharyngeal factors in maintaining drinking in these species is thus clear, there has been very little evidence on the applicability of this conclusion to non-human primates such as monkeys, and to man. Therefore it is of interest to consider the results of recent studies of sham drinking in the rhesus monkey (Maddison, Rolls, Rolls & Wood, 1977; Maddison, Wood et al., 1980). The

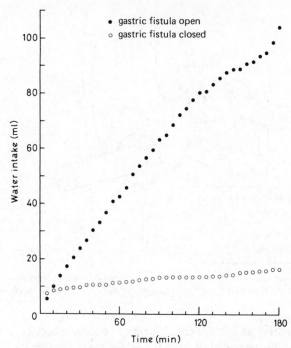

Fig. 5.1. Sham drinking by a rat with an open gastric fistula to allow ingested water to drain out, compared with normal drinking with the stomach fistula closed. Drinking is maintained (and in fact there is overdrinking) in the sham-drinking condition, indicating that the maintenance of drinking depends on oropharyngeal and related factors, and not in the first instance on dilution or expansion of the body fluids.

monkeys were equipped with a gastric cannula and a duodenal cannula relatively close to the pylorus (see figure 5.2), both of which were normally closed. When both cannulae were closed, the monkeys deprived of water for 22.5 h drank approximately 135 ml in the 1-h test session (figure 5.3). When the gastric or the duodenal cannula was open so that water did not reach the intestine and was not absorbed, drinking still occurred (figure 5.3). (In fact, there was much more drinking than with the cannulae closed – see Chapter 6.) Thus, the monkeys did drink when oropharyngeal factors such as taste were provided by the water, but when no water was absorbed or entered the intestine. This shows that in the primate, as well as in the dog, horse and rat, the animal works to obtain oropharyngeal and other sensations provided by drinking the water, and that the gradual dilution of plasma or expansion of extracellular fluid volume produced by the ingestion of water is not necessary for the maintenance of drinking.

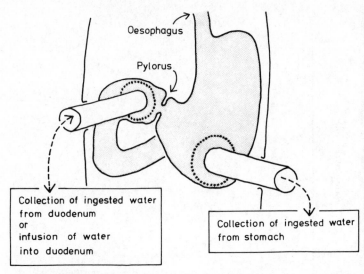

Fig. 5.2. Diagram showing the cannulated preparation used in monkeys for the investigation of peripheral factors in drinking. (From B. J. Rolls, Wood & Rolls, 1980.)

It thus appears that animals normally drink to obtain oropharyngeal and other sensations provided by water, and not (in the first instance) because body fluid deficits are being removed. This view is strengthened by the evidence that animals have difficulty in learning to self-administer water intragastrically (Epstein, 1960) or intravenously (Nicolaïdis & Rowland, 1974). For example, relatively large volumes (e.g. normally 1 ml, but in the range 0.25–2 ml) of water must usually be delivered for every bar press if the animals are to learn to press a bar to self-administer water that does not pass through the oropharynx. In contrast, if an animal can taste the water, then it readily learns to perform a response to obtain a much smaller volume of water. It is in this sense that oropharyngeal factors guide and maintain (or provide the reward or incentive for) water intake normally. This view is in line with the fact that oropharyngeal receptors, e.g. taste receptors, are specialized in the rapid and sensitive detection of small quantities of substances which may be rewarding, such as water for the thirsty animal. In the course of evolution, or perhaps as a result of learning, animals must have adapted to work for sensations such as the taste of water when they are thirsty, i.e. for sensations which precede the removal of fluid deficits by the ingested water. This clearly provides a more sensitive and more rapid mechanism to guide behaviour than that which could be provided by changes in systemic fluid balance. Another selective

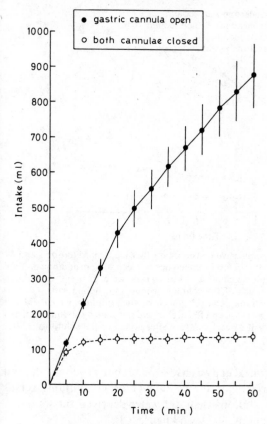

Fig. 5.3. Cumulative water intake in five monkeys after 22.5 h water deprivation, with the gastric and duodenal cannulae both closed and with the gastric cannula open. Means ($\pm$S.E.) are shown. (From Maddison, Wood, Rolls, Rolls & Gibbs, 1980.)

advantage of this peripheral, sensitive, pre-absorptive control over whether drinking is maintained is that it allows behaviour to be guided by previous experience with a substance, without requiring significant ingestion of the substance. This is important, for example, in learned aversion (or bait shyness), in which an animal learns to reject the taste of a solution which has previously been associated with sickness (see Chapter 7) (B. J. Rolls & Rolls, 1973; Garcia, Hankins & Rusiniak, 1974; Barker, Best & Domjan, 1977). Another advantage of this pre-absorptive, oropharyngeal, control of drinking is that it allows drinking to be guided to homeostatically relevant solutions. An example of this is that rats which had been on a low-sodium diet (or in which

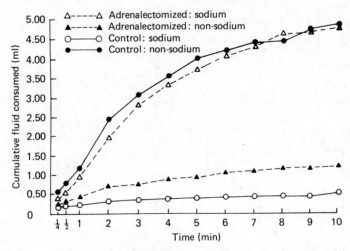

Fig. 5.4. Mean cumulative consumption by 41 adrenalectomized and 41 control rats in a 10-min test of preference between sodium and non-sodium solutions. The adrenalectomized rats, which were deficient in sodium, drank relatively much of the solutions containing sodium salts, whereas the control rats drank relatively much of the solutions which did not contain sodium salts. This preference for sodium salts by sodium-depleted animals was evident as early as 15 s after access to the solutions. (After Nachman, 1962.)

adrenalectomy or peritoneal dialysis had lowered body sodium) showed a relatively greater preference for sodium chloride solutions versus water or non-sodium solutions than did water-deprived rats (see e.g. Nachman, 1962; Fitzsimons, 1979; and figure 5.4).

How oropharyngeal factors may maintain drinking

The experiments just described show that normally oropharyngeal stimulation guides water intake (so that when they are deprived of water, animals ingest water rather than food or hypertonic saline, for example) and maintains intake. Oropharyngeal stimulation thus provides the incentive, or in this sense the reward, for drinking in the thirsty animal. (For a further discussion of this see Rolls, E. T., 1975).

A reward may be defined as anything for which an animal will work. The subjective sensations underlying the reward value that water has for the thirsty animal have only recently been investigated experimentally. Human subjects were asked to give a rating of how pleasant water tasted after overnight water deprivation, and during as well as after drinking. The rating was made by the subjects marking a point, corresponding to

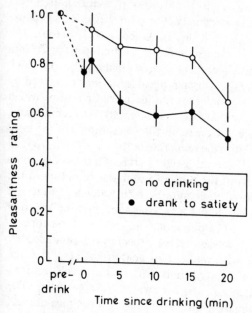

Fig. 5.5. The pleasantness of the taste of water in relation to drinking. Human subjects deprived of water overnight (until testing at 11 a.m.) rated the pleasantness of the taste of water on a visual analogue scale (the subjects marked the subjective pleasantness of the taste on a 10-cm line labelled at one end 'very pleasant' and at the other 'very unpleasant'). This rating is shown as the 'predrink' rating of 1 on the graph. Then one group of 20 subjects was allowed to drink water to satiety. Their ratings of the pleasantness of the taste of small samples of water at different times after the drinking are shown and it is clear that the taste of water became less pleasant following drinking to satiety ($P<0.001$). The ratings at each time are expressed as a ratio of the predrink rating, and the means ($\pm$S.E.) are shown. In a second group of 12 subjects, no drinking of the water to satiety was allowed, but ratings of the pleasantness of the taste of small samples of water at different times after the start of testing were made for comparison with the control group. In this second group, being allowed to taste the water without ingestion did make water taste a little less pleasant, but the effect was smaller than in the group allowed to drink the water ($P<0.025$ for the difference between the two groups.) (From experiments by E. T. Rolls, B. J. Rolls & J. Seton.)

the pleasantness of the water in a 30-s rinsing period, on a visual analogue scale, in this case a 10-cm line labelled at one end 'very pleasant' and at the other 'very unpleasant'. It was found that the subjects in the water-deprived state gave a high rating to the pleasantness of the taste of water immediately before drinking. After the subjects had drunk to satiety, a state reached after drinking for 2–7 min, the taste of water was considerably less pleasant than it was initially, when the subjects were thirsty (the ratio was 0.76; see figure 5.5, from experiments by E. T. Rolls, B. J. Rolls & J. Seton). The pleasantness of the taste of water continued to diminish for 20 min following the drinking. This experiment suggests that one way in which thirst stimuli act is to make the oropharyngeal sensations produced by water pleasant, and that this underlies the incentive or reward which water has for the thirsty animal. Also, it suggests that, conversely, satiety mechanisms produce a decrease in the pleasantness of the taste of water, and that this underlies the reduction in the incentive or reward that water has as the animal becomes satiated. (It is of interest that some decrease in the pleasantness of the taste of water was found when the human subjects simply rinsed their mouths with water and were not allowed to drink to satiety (see figure 5.5). Thus some minor alleviation of thirst by oropharyngeal stimulation alone can occur.) Thirst appears to modulate the subjective sensations and the reward produced by oropharyngeal stimulation with water. This analysis suggests that the subjective phenomenon just described is part of the mechanism by which thirst signals influence whether water is consumed. The phenomenon is perhaps analogous to the decrease in the pleasantness of the smell or taste of food produced in human subjects by feeding (Cabanac, 1971; E. T. Rolls & Rolls, 1981; B. J. Rolls, Rolls, & Rowe, 1981), and termed by Cabanac (1971) alliesthesia, that is literally, 'changing sensation'.

The finding that thirst modulates the pleasantness of the taste of water has been investigated in a second experiment in man, in which estimates of the hydration of the cellular and extracellular fluid compartments were made to allow direct comparison with the subjective sensations. It was found that the rating (again on a visual analogue scale) of 'how pleasant would it be to drink some water now' was elevated significantly (compared to the non-deprived state) towards pleasantness after 24 h of water deprivation (figure 1.2). Also, the pleasantness returned rapidly towards the baseline non-deprived value in the 5–10-min period in which most of the drinking occurred (figure 1.2) (B. J. Rolls, Wood, Rolls *et al.*, 1980). The increase in the pleasantness rating was associated with the depletions of the cellular and extracellular compartments shown in figure 6.3 (see next chapter) and with willingness to drink water when it was provided. Comparison of figure 1.2 and figure 1.3 shows that the subjective pleasantness of the taste of water decreased at the same time as the rate of drinking decreased (B. J. Rolls, Wood, Rolls *et al.*, 1980). Thus, the sensation of the pleasantness of the taste of water could be a subjective part of

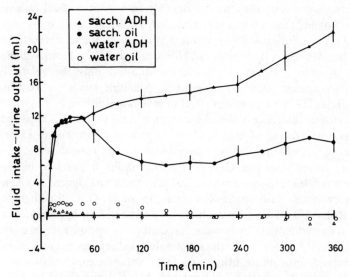

Fig. 5.6. Mean fluid balance (fluid intake minus urine output) of 12 rats drinking either saccharin (sacch.) solution or water after an injection of antidiuretic hormone (ADH) or oil vehicle. The rats were in fluid balance at the start of the experiment. The palatability of the saccharin provided a powerful incentive, and induced so much drinking that if excretion of water was impaired with ADH, the rats went markedly and dangerously into positive fluid balance. (From B. J. Rolls, Wood & Stevens, 1978.)

the mechanism by which the acceptability of water, and drinking, is controlled.

Effects of palatability on drinking and body fluid homeostasis

The experiments described above show that thirsty animals normally work in order to obtain oropharyngeal sensations such as the taste of water. The importance of oropharyngeal sensations in maintaining or providing the incentive for fluid intake is seen very dramatically in experiments in which the palatability of the fluid is enhanced. For example, making the fluid taste sweet by the addition of saccharin makes it so rewarding that even non-deprived rats drink large quantities of it (Ernits & Corbit, 1973; B. J. Rolls, Wood & Stevens, 1978). This effect is seen to be very powerful in a clinically relevant paradigm (B. J. Rolls, Wood & Stevens, 1978), in which the ability of rats to excrete water by producing dilute urine is impaired by the administration of antidiuretic hormone. When the rats, which were in fluid balance at the start of the experiment, were offered saccharin solution, they drank so much that they went markedly into positive fluid balance (figure 5.6) and their plasma was diluted by an average of 22 mosmoles/ kg H_2O. The dilution was so extreme that in some animals it resulted in some haemolysis or rupture of red blood cells due to osmotic swelling. Thus, inhibitory signals from plasma dilution must be relatively ineffective against the consumption of a palatable saccharin solution, and palatibility has a powerful effect in inducing drinking.

The clinical relevance of this observation is that if excretion is impaired, for example because of high levels of antidiuretic hormone or kidney failure, polydipsia may cause plasma dilution and hyponatraemic (low sodium) convulsions (see Chapter 9). Palatability appears to be one factor which leads to overdrinking and can exacerbate clinical problems of inappropriate fluid intake. For example, we (J. G. G. Ledingham, B. J. Rolls & J. Gibbs) have observed a patient with hyponatraemic convulsions who drank little water but consumed up to 100 cups of tea a day (over 15 l). Normally the mammalian kidney can excrete excess fluid rapidly in a dilute urine, and large volumes can be consumed without seriously affecting fluid balance. But if excretion of water by the kidneys is impaired, the effect of palatability may be so powerful that it leads to fluid intake even in the presence of plasma dilution, with pathological results.

Effects of variety on fluid intake

We have just seen that giving liquids a pleasant taste significantly increases fluid consumption. Another factor unrelated to physiological need which may affect intake is the variety of fluids available. It has been shown that offering foods with a variety of flavours significantly increases feeding (Le Magnen, 1956; B. J. Rolls, 1979; B. J. Rolls, Rolls & Rowe, 1981; B. J. Rolls, Rowe & Rolls, 1981; B. J. Rolls, Rowe, Rolls *et al.*, 1981). To determine whether this effect is found for drinking, we offered rats that were in fluid balance either water for 1 h, water with one of four artificial essences added (orange, raspberry, malt and peach, supplied by Barnett and Foster) for 1 h, or water with a different essence every 15 min over a 1-h period (i.e. they had orange, raspberry, malt and peach in succession over 1 h). It can be seen in figure 5.7 that offering rats a variety of fluids to consume significantly increased intake (B. J. Rolls, Wood & Rolls, 1980).

Variety can also increase fluid consumption in humans (B. J. Rolls, Wood & Rolls, 1980). Non-deprived subjects, under the pretext of a tasting experiment, consumed three drinks successively, with a 10-min

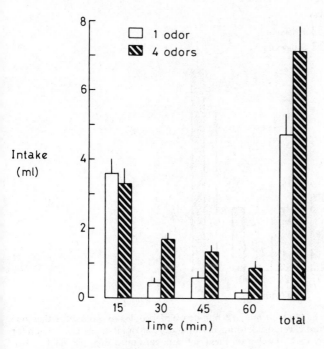

Fig. 5.7. Over 1 h, 15 non-deprived rats were offered either water with one artificial essence added (orange, raspberry, peach or malt), or all four essences in succession for 15 min each. These rats had consumed 2.5 ± 0.4 ml plain water in the same situation. Adding just one essence to water increased intake by 88% ($P<0.001$) compared with plain water and changing the smell every 15 min over 1 h increased intake by 182% ($P<0.001$) compared with plain water and by 50% ($P<0.001$) compared with just one essence. Means ($\pm$S.E.) are shown. (From B. J. Rolls, Wood & Rolls, 1980.) Thus variety can enhance fluid intake.

period allowed for each drink, under three different conditions: three different flavours (low-calorie orange, lemon and lime drinks), one flavour only, or water alone with no flavour. More was consumed in the three-flavour than the one-flavour condition, and more in the one-flavour than in the no-flavour condition (see figure 5.8). Thus, the experiments in both rat and man show that offering a variety of fluids over a short period stimulates significantly more intake than is needed for homeostatic control. These experiments provide another demonstration of how oropharyngeal sensations are important in providing the reinforcement for fluid intake.

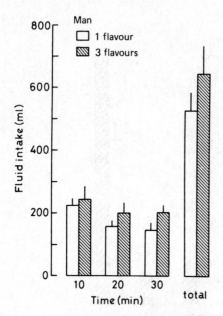

Fig. 5.8. Over 30 min, 18 non-deprived people were offered either one low-calorie fruit drink (orange, lemon or lime) or all three in succession for 10 min each. Twelve of these subjects consumed 265 ± 68 ml of plain water when tested in the same situation. Adding just one flavour to water increased intake by 99% ($P < 0.001$) compared with water, and giving three different flavours increased intake by 143% ($P < 0.001$) compared with plain water and by 22% ($P < 0.001$) compared with just one flavour. Means ($\pm$S.E.) are shown. (From B. J. Rolls, Wood & Rolls, 1980). This shows that palatability and variety can enhance fluid intake.

Conclusions

Experiments with sham drinking show that fluid intake is guided and maintained by oropharyngeal sensations. This allows for rapid control of behaviour before the absorption of ingested fluid has had time to occur. Although animals work to obtain oropharyngeal sensations, and not to sense immediately plasma dilution or expansion, water intake is normally, as a result of evolution and of learning, appropriate for homeostatic need (see Chapter 10). The operation of this mechanism implies that the rewarding value of oropharyngeal stimulation by water depends on the presence of thirst stimuli, so that the taste of water is only rewarding in the thirsty animal. Some evidence for the operation of this mechanism is available from subjective reports made by humans. They rate the taste of water as being more pleasant when they are thirsty than when they are satiated. More work needs to be done to investigate how this mechanism operates. Another advantage of this pre-absorptive mechanism for guiding intake is that it allows the learning of preferences for or aversions to fluids with particular tastes based on the previously experienced consequences of their ingestion. For example, learned taste aversion can allow the ingestion and absorption of potentially dangerous substances to be avoided. The powerful role of oropharyngeal sensations in guiding and maintaining intake is seen in experiments in which palatability and variety can lead to enhanced fluid intake (which can even be pathological if excretion is impaired).

6   The termination of drinking

Introduction

Body fluid deficits such as cellular dehydration or depletion of extra-cellular fluid volume which lead to the initiation of drinking have been described in Chapter 4. But what stops drinking? One possibility is that drinking continues until the body fluid deficits that initiated the drinking are repleted (figure 6.1e). But because absorption of ingested water must take time, this mechanism is unlikely to be sufficiently rapid to account for the termination of drinking, except perhaps in species in which the drinking pattern is slow and intermittent. This suggests that a pre-absorptive mechanism using signals from the mouth (figure 6.1a), the oesophagus (figure 6.1a), the stomach (figure 6.1b), or the duodenum (figure 6.1c) could be important in satiation. For example, gastric distension (Fig. 6.1b) could provide information on the volume of fluid ingested, and duodenal osmoreceptors (figure 6.1c) on the tonicity of the fluid. Another possibility is that receptors in the hepatic–portal system (figure 6.1d), which conveys absorbed materials from the gut to

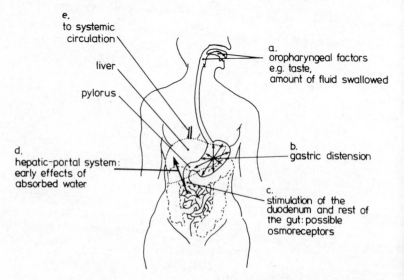

Fig. 6.1. Some of the possibilities for signals which terminate drinking are illustrated (see text).

the liver, or receptors in the liver itself may signal that dilution is occurring before there have been large changes in the systemic (i.e. general) circulation. Some of these possibilities are illustrated in figure 6.1.

Ways in which the contributions of these different factors can be separated and quantified are described below. At this stage it could be noted that the termination of drinking is under relatively precise control. For example, in a 1-h drinking test following 24 h water deprivation, dogs and man drink relatively accurately the volume of water required to restore the fluid deficits (see figure 6.3 below) and do not usually overdrink or produce a great deal of dilute urine (see below and B. J. Rolls & Wood, 1977b; and B. J. Rolls, Wood, Rolls *et al.*, 1980)*. If overdilution of the body fluids occurs it can be dangerous, leading to symptoms of water intoxication such as nausea, vomiting and confusion (see Chapter 9).

An important preliminary point is that satiation or the termination of drinking is more difficult to define behaviourally than is the onset of drinking. This follows from the drinking pattern, which has a clearly defined start but, particularly in the rat for example, has a much less clearly defined termination (see figure 6.2). Thus, care is necessary in specifying a criterion for the termination of drinking. Also, drinking is easily disrupted by experimental manipulation, so that if an animal does not drink it is difficult to know whether it is no longer thirsty or is in some way upset by an experimental regime. The observation that there is a satiety sequence for drinking in the monkey, in which a period of vigorous drinking is followed by a period of sleepiness (Maddison, Wood *et al.*, 1981) and the analysis of subjective reports as satiety is reached in man (B. J. Rolls, Wood, Rolls *et al.*, 1980; see below) are helpful in determining true satiety as compared with interference with drinking.

The pattern of drinking

In attempting to define the crucial events for stopping drinking it is particularly informative to examine the pattern of drinking in a

---

* With more severe water deprivation, drinking appears to undercompensate (Adolph, 1950), but this may reflect in part the voluntary undereating which accompanies severe water deprivation, so that it should not be expected that lost body weight would be made up by drinking alone. This emphasizes the value of taking measures of body fluid composition, rather than just estimating deficits from body weight loss. Another factor in the apparent undercompensation is that species that drink slowly will only have completed a part of their drinking in a short, 15-min test period.

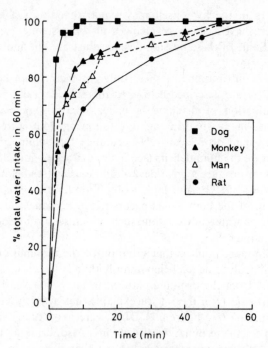

Fig. 6.2. Water intake plotted as a percentage of the total intake in 1 h after overnight water but not food deprivation. The dog is a rapid drinker, completing its drinking within 10 min, whereas the rat drinks comparatively slowly. The monkey and man fall between these extremes. (From B. J. Rolls, Wood & Rolls, 1980.)

particular species for clues about the possible mechanisms which terminate drinking in that species. The pattern of water intake in four different species is shown in figure 6.2. The intake was measured after a standard 21–24 h overnight water-deprivation period. The dog is a very rapid drinker and typically consumes almost all of its requirements within 2–3 min of the onset of drinking (Adolph, 1939). Some form of peripheral metering thus seems a probable mechanism for the termination of drinking in the dog.

Systemic factors in the termination of drinking

In order to assess the importance of changes in systemic body fluid levels as a result of water absorption in satiety, it is necessary to know how rapidly these changes occur after access to water and whether the body fluid changes that occur are sufficient to terminate drinking. Ways in which these questions can be answered are described in this section.

How rapid is the restoration by drinking of body fluid deficits?

First, the body fluid changes which actually occur during the drinking illustrated in figure 6.2 following 21–24 h overnight water deprivation are shown in figure 6.3 for the dog, monkey and man. The measurements

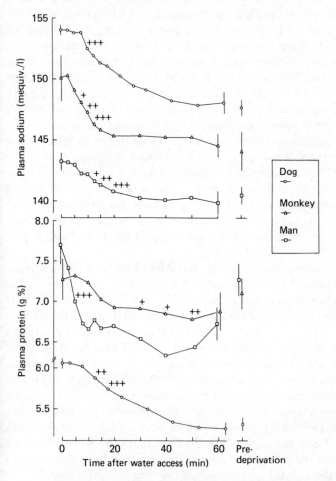

Fig. 6.3. Changes in plasma sodium concentration and plasma protein concentration during and following drinking in 21–24h water-deprived dogs (circles; $N = 8$), monkeys, (triangles; $N = 5$) and humans (squares; $N = 5$). The subjects were allowed access to water at the start of the hour. Means ($\pm$S.E.) are shown. Significant differences from initial values: $+$, $P<0.05$; $++$, $P<0.01$; $+++$, $P<0.001$. Prepared from data of Ramsay, Rolls & Wood, 1977a (dog), Wood *et al.*, 1980 (monkey), and B. J. Rolls, Wood, Rolls *et al.*, 1980 (man).

were made by taking a plasma sample before the deprivation started and repeated samples during the 1-h drinking period. The extent of cellular dehydration is reflected by plasma sodium concentration (and by the plasma osmolality, which was also measured in the original studies but is omitted from figure 6.3 for the sake of simplicity as the agreement was very good); and by the extent of plasma volume depletion by plasma protein concentration (and by haematocrit in the original studies). For the dog, comparison of figure 6.2 and figure 6.3 shows that drinking terminated within 2–3 min after access to water, whereas plasma dilution and expansion did not become significant until 10–12 min after access to water and did not reach predeprivation levels (shown at the right of figure 6.3) until 40–45 min after access to water (Ramsay, Rolls & Wood, 1977a). Thus, in this rapid drinker systemic signals reflecting cellular rehydration and extracellular fluid expansion cannot account for the termination of drinking, and peripheral, presystemic, factors must be crucial. Although most of the dog's drinking is completed within 2–3 min, the accuracy of the rehydration is clear from the final values of the plasma sodium and protein concentrations, which approached the predeprivation values very closely, and from the fact that there was no excretion of dilute urine, so that overdrinking had not occurred (B. J. Rolls & Wood, 1977b).

In the water-deprived monkey, the initial bout of relatively rapid drinking typically lasted for about 7.5 min. After that time, further intake was smaller and more sporadic (figure 6.2). Following the start of drinking, plasma sodium concentration fell rapidly, the change becoming significant after 7.5 min (figure 6.3). Plasma protein concentration fell appreciably during the first 20 min, but this change did not become statistically significant until 30 min after access to water. Thus, in the monkey the absorption of water is relatively rapid and cellular dehydration (as assessed by plasma sodium concentration) as a thirst stimulus is starting to change at the time when drinking is being attenuated. However, it takes 20 min or more for the major change in both the cellular and extracellular fluid compartments to be completed (figure 6.3), so that some factor other than systemic body fluid changes may contribute to the termination of drinking in the monkey (Wood *et al.*, 1980). Some signal, such as pre-absorptive metering, may be required to restrain drinking for the period 7.5–30 min after the onset of drinking.

It is clear from a comparison of figures 6.2 and 6.3 that man is a relatively rapid drinker, drinking much of his total intake in the first 2.5 min following 24 h water deprivation, but continuing to ingest small volumes of water during the remainder of a 1-h test; and that plasma dilution (shown by osmolality in addition to the sodium concentration changes illustrated) associated with this drinking has a slower time-course. Some dilution is apparent 7.5 min after access to water, but consistent and significant changes are not apparent until about 12.5 mins after access to water (figure 6.3) (B. J. Rolls, Wood, Rolls *et al.*, 1980). Thus, replenishment of the cellular fluid deficit cannot account for the termination of drinking in man, although by 12.5–20 min it is probably an important factor in limiting further drinking. The change in plasma protein concentration following drinking was rapid in this experiment and may have resulted from food being present in the gut, so that drinking water produced an isotonic fluid which was rapidly absorbed, or from a shift of extracellular fluid into the vasculature, as well as from absorption of some water. This rapid hypervolaemia in man is unlikely to be a major factor in the termination of drinking in that only a small (5%) reduction in drinking was produced in the monkey by infusions of isotonic sodium chloride which produced a marked plasma hypervolaemia (see below). A similar experiment in man would of course be interesting. It may be concluded that, in man, cellular rehydration does not appear to be sufficiently rapid to account for the early decrease in the rate of drinking, but may contribute to satiety, later on, after 12.5 min. Expansion of plasma volume, although rapid, may not be a powerful factor in terminating drinking. Therefore, a contribution of other, presystemic, factors to the early termination of drinking must be suspected.

During these experiments on the control of drinking in man, it was possible to take subjective ratings of thirst, the pleasantness of the taste of water, the dryness of the mouth, etc., on a visual analogue scale. It is clear that during drinking to satiety, these ratings altered rapidly (figure 1.2) before much plasma dilution or expansion had occurred (figure 6.3), so that presystemic factors must be important in producing subjective changes, including the human subjective feeling of satiety (see figure 1.2).

In the rat it has been difficult to study possible systemic factors involved in the termination of drinking because the small size of the animal has made continuous assessment of plasma composition difficult (Rolls & Wood, 1979). If a single plasma sample is taken at the end of drinking it is found that a significant reduction in plasma osmolality (or sodium) and even overdilution has occurred while plasma volume is still depleted (figure 6.4) (Hatton & Bennett, 1970; Hall & Blass, 1975; Blass & Hall, 1976). Thus, it appears that in the rat drinking proceeds at a rate which is slow enough for significant

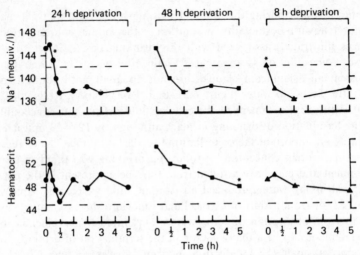

Fig. 6.4. Mean serum sodium and mean haematocrit of rats drinking water after 24, 48, or 8 h of water deprivation. Asterisks indicate values that do not differ statistically from *ad libitum* values (broken line). Serum sodium estimates cellular dehydration; haematocrit estimates extracellular fluid volume. (From Blass & Hall, 1976.)

absorption to take place before drinking terminates. The reversal of the cellular fluid deficit may thus be an important factor in the termination of drinking in this species, which drinks relatively slowly.

How effective are systemic changes in producing satiety?

A second line of evidence on the role of systemic dilution and expansion in satiety comes from experiments in which the effects of systemic infusions of water or saline on the termination of drinking are measured. Do these infusions stop drinking? In one type of study, it is found that continuous intravenous infusions of water in the rat only reduce, and do not eliminate, drinking. This is found even when the rate is increased at meal times when the rat would normally drink. In particular, premeal and non-meal associated drinking persist at about 5 ml/24 h (Nicolaïdis & Rowland, 1975b; Rowland & Nicolaïdis, 1976; Rowland, 1977). (The normal intake of the rats was approximately 30 ml/24 h.) This type of study thus suggests that systemic factors alone are sufficiently powerful (if present) to cause a large reduction in drinking, but not to prevent it completely.

In another type of study, infusions are given to eliminate the systemic changes produced by water deprivation, and the effect this has on the drinking is measured. In the rat, it was found with this design that the drinking following 24 h water deprivation was reduced by 64–69% by preloads of water which reversed the cellular dehydration produced by the deprivation, and that the drinking was reduced by 20–26% by preloads of isotonic saline which reversed the extracellular fluid depletion produced by the deprivation (Ramsay, Rolls & Wood, 1977b). Thus, in the rat reversal of the cellular fluid deficit, which occurs as this relatively slow drinker terminates drinking (Hatton & Bennett, 1970), is probably a factor (but cannot be the only factor – see above) in terminating the drinking.

In the dog, drinking following 24 h water deprivation was reduced by 72% when osmolality was reduced to the predeprivation level (Ramsay, Rolls & Wood, 1977a). In this case, only the osmolality of the blood supply to the head was reduced, by giving the infusion into the intracarotid artery (and monitoring its effect via the blood leaving the head in the jugular vein), so that in this case it can be concluded that systemic dilution sensed centrally, in the brain, can reduce deprivation-induced drinking by 72%. Expansion of the extracellular fluid in the dog produced by an intravenous infusion of isotonic saline reduced the deprivation-induced drinking by 27%. A combination of both procedures, thus removing the centrally sensed cellular dehydration and the extracellular fluid depletion, reduced drinking to a very low value in the dog, so that the removal of the cellular and extracellular stimuli for drinking following water deprivation was sufficient to abolish drinking. However, although in the dog the reversal of the sensed fluid deficits can abolish drinking following deprivation, it has been shown above that the reversal of the deficits during normal drinking is not sufficiently rapid to account for the normal termination of drinking which occurs in the dog within 2–3 min after access to water. There must be some additional mechanism, at least to account for the initial termination of drinking, although after approximately 40–45 min fluid repletion is sufficient to explain the inhibition of further drinking.

In the monkey, it has been possible with intravenous infusions of water to mimic the decrease in systemic osmolality which occurs by absorption of ingested water during normal drinking following water deprivation (Wood *et al.*, 1980). This has been done in monkeys which were sham drinking while ingested water was drained from the stomach. It was found that the sham drinking was reduced by the intravenous infusion of water (100 ml), but that drinking did not stop, as it would if the systemic dilution normally occurring during drinking was a sufficient factor to terminate drinking (figure 6.5). In fact, this experiment shows that even the combination of systemic dilution and oropharyngeal stimulation produced by the sham drinking, for example by the taste of

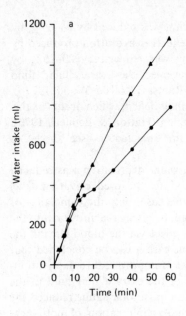

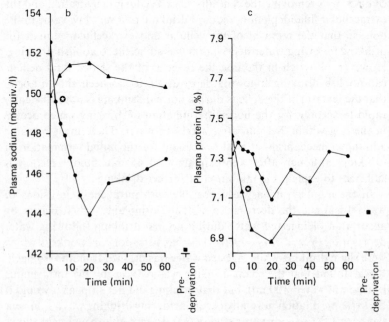

Fig. 6.5. The effect of 100 ml intravenous infusions of water (circles) or isotonic saline (triangles) on (a) drinking, (b) plasma sodium concentration, and (c) plasma protein concentration in the monkey. The infusions were

water and swallowing, are not sufficient to account for the termination of drinking. Some other factor, such as gastric distension, stimulation of the gut by water, or expansion of the extracellular fluid, must also be necessary. These other factors will be considered below, but it can be noted here that since expanding plasma volume (with 100 ml intravenous isotonic saline) to a similar degree as occurs during normal drinking did not reduce gastric sham drinking, plasma expansion is unlikely to be an important factor in producing satiety in the monkey (figure 6.5; Wood *et al.*, 1980).

In conclusion, measurements of systemic dilution and expansion show that these processes do not generally occur rapidly enough to account for the termination of drinking. This is the case in rapid drinkers, such as the dog, and in relatively rapid drinkers, such as the monkey and man. In slow drinkers, such as the rat, sufficient absorption to restore at least partly the body fluid deficits has occurred by the time drinking terminates, and this type of systemic change has been shown to contribute to the termination of drinking. Removal of the body fluid deficits by absorption approaches completion 30–50 min after access to water, and is thus gradually able to contribute to the continuing suppression of drinking in rapid drinkers. Particularly important in this suppression of drinking is removal of the cellular stimulus to thirst, as shown by infusion experiments.

Oropharyngeal factors in the termination of drinking

Given that systemic factors do not change rapidly enough to account for the termination of drinking in many species, some other, rapid, perhaps pre-absorptive, factor or factors must be important. One such possible mechanism could arise from the oropharyngeal stimulation which occurs while water is being ingested. Simply tasting the water and swallowing it could be sufficient to induce satiety and terminate drinking. How can this be tested? A direct approach is to consider what happens to drinking when ingested water is allowed to drain from an oesophageal or gastric fistula, that is, in oesophageal or gastric sham

given at 5 ml/min starting at time 0, the time when sham drinking, with drainage from the stomach, was allowed to start. The infusions were designed to mimic either the plasma dilution or the plasma expansion which occur during normal drinking. Systemic dilution with water reduced the drinking but did not stop it. Plasma expansion with saline had very little effect on drinking in the monkey. Times at which the plasma variables first differ significantly from the initial values during the infusions are indicated by asterisks, $P < 0.05$. Predeprivation levels are indicated by closed squares. (Modified from Wood *et al.*, 1980.)

drinking (see figure 3.4). With these procedures, water can be tasted and swallowed normally, yet does not accumulate in the stomach causing gastric distension, nor pass into the gut to cause systemic dilution following absorption.

With sham drinking, it is found that drinking is not terminated normally and that overdrinking occurs. The failure to terminate drinking normally under sham-drinking conditions is made quite clear in figure 5.1, in which a rat continued to drink for 3 h when water drained from a gastric fistula. More than 100 ml was swallowed, compared to 10–15 ml in the same period when the fistula was closed. This overdrinking under sham-drinking conditions has been described in the horse (Bernard, 1856), the dog (Bernard, 1856; Bellows, 1939; Towbin, 1949; Adolph, 1950), the rat (Blass, Jobaris & Hall, 1976), the sheep (Bott, Denton & Weller, 1965) and more recently in monkeys, which have also been shown to overdrink greatly when a gastric cannula is opened to allow drainage of ingested water (see figure 5.3; Maddison, Wood *et al.*, 1980). It should be noted that the over-drinking occurred the first time the cannula was opened, so that it cannot be argued that the overdrinking is a result of learning. The conclusion which follows is that in most species oropharyngeal stimulation by water is not sufficient to account for the termination of drinking.

Nevertheless, there is some evidence that oropharyngeal stimulation can make a contribution to satiety and the termination of drinking. Part of the evidence comes from the early studies on sham drinking in the dog, in which it was found that, although overdrinking occurred, the amount sham drunk was proportional to the magnitude of the water deficit (Bellows, 1939; Towbin, 1949). Often, after the dogs had drunk approximately 2.5 times as much water as the intact dog under the same deprivation conditions, the sham drinking stopped, at least temporarily. In the rat, it has also been found that more sham drinking occurs when thirst is made more extreme by longer water deprivation (Blass, Jobaris & Hall, 1976). These findings have been taken to imply that a contribution to the termination of drinking is made by oropharyngeal metering of the amount of water consumed, so that the thirstier the animal is, the more water is sham drunk, but eventually some inhibition of drinking occurs.

Other evidence that oropharyngeal stimulation makes a contribution to the termination of drinking comes from an experiment by Miller, Sampliner & Woodrow (1957). There was less drinking after water deprivation if a 14-ml preload of water, given before the drinking test, was consumed by the rats by mouth than when the 14-ml preload was placed directly into the stomach through a fistula (figure 6.6). This

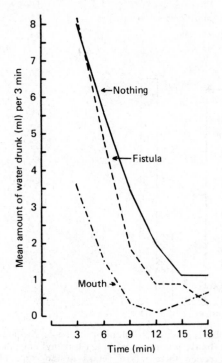

Fig. 6.6. The effect of 14-ml preloads of water delivered via the mouth or stomach on water intake of rats during the 18 min immediately after the loads. It can be seen that water consumed normally is more effective than that tubed into the stomach. (From Miller, Sampliner & Woodrow, 1957.)

suggested that the oropharyngeal stimulation and related factors contribute to the termination of drinking. This contribution of oropharyngeal factors is probably particularly important in the short term, in the early stages of satiety, and was prominent in the first 15 min after access to water in this experiment (see figure 6.6). In the longer term, other, post-absorptive, factors become more important, as shown by the observation that oral, intragastric and intravenous routes of administration of a 10-ml preload of water (or saline) produced equal reductions in water intake after water deprivation in the rat, when this was measured over a 1-h period (figure 6.7) (analysis of data from Ramsay, Rolls & Wood, 1977b).

Given that oropharyngeal stimulation makes a contribution to the termination of drinking, though is not sufficient when acting alone to account for the normal termination of drinking in most species, it may be asked to what extent oropharyngeal stimulation produces subjective

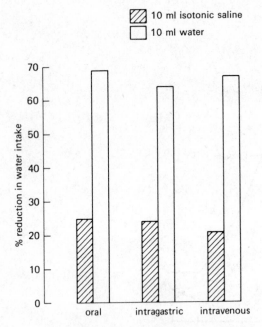

Fig. 6.7. The effect of preloads of water or isotonic saline on deprivation-induced drinking by rats. Intake was measured in a 1-h period after the preloads. Over this 1-h period there was no effect of the route of administration of the preload on total intake. (After Ramsay, Rolls & Wood, 1977b.)

feelings of thirst reduction and satiety in man. There is some evidence available on the satiating effects of oral stimulation by water in man (B. J. Rolls, Wood, Rolls *et al.*, 1980). In human subjects deprived of water for 24 h, rinsing the mouth for 30 s with water produced a relief of thirst, which was however only very temporary, outlasting the stimulation by only 30–50 s. This is in line with the evidence that even combined oral and pharyngeal stimulation by water (i.e. including swallowing) does not produce sufficient satiety to terminate drinking normally in most other species, although it does have some effect in inhibiting drinking. A greater inhibition of thirst may be produced by oral stimulation with cold water (see Chapter 10 and figure 10.2). For example rats drink 1.5–1.7 times more water at body temperature than water at 12 °C (Kapatos & Gold, 1972a, b; Gold *et al.*, 1973).

In conclusion, sham-drinking experiments show that oropharyngeal factors acting alone are not sufficient to account for the normal termination of drinking in most species, but oropharyngeal metering may be able to make a rapidly acting contribution to the termination of drinking. The evidence implying that oropharyngeal factors can contribute to satiety is confirmed by observations that oral stimulation by water in man can produce feelings of satiety even if these are only short-lasting.

Gastric factors in the termination of drinking

As systemic factors do not in general act rapidly enough to terminate drinking, and oropharyngeal metering, although potentially a rapidly acting mechanism, is not sufficient alone to produce satiety, it becomes a possibility that gastric factors, for example gastric distension, which would act rapidly, at least contribute to the early termination of drinking. There are several ways in which the contribution of gastric factors can be analysed and these are discussed next.

Gastric distension

First, a water load can be introduced into the stomach immediately before drinking. This attenuates the drinking in some species, for example in the rat, hamster and guinea pig, although it is less effective in the dog (Adolph, 1950; see also Sobocińska, 1978; Ramsay & Thrasher, 1980). The effect is probably produced by mechanical distension of the stomach, for it can also be produced by intragastric injection of sea water or air (Adolph, 1950), or by the inflation of a balloon in the stomach (Towbin, 1949; Adolph, Barker & Hoy, 1954). A possible pathway for information about gastric stretch to reach the brain is via the vagus nerve. However, the effect of vagotomy is not simple to interpret, for in the rat bilateral abdominal vagotomy tended to decrease drinking induced by water deprivation (it also reduced drinking in response to cellular dehydration and decreased drinking induced by polyethylene glycol; see Kraly, Gibbs & Smith, 1975; Kraly, 1978). The gastric distension experiments themselves are not easily interpreted because it is not clear whether the exogenously administered gastric load mimics the distension which normally occurs at the termination of drinking. In particular, distension with a balloon may be painful, and thus aversive, and for this reason inhibit drinking. For example, Miller (see Grossman, 1967, p. 434) has found that rats will avoid a response followed by gastric distension. Another problem with the gastric loading experiments is that some of the intubated water will have left the stomach and may contribute to satiety by post-gastric mechanisms. Methods of investigation which avoid some of these problems are presented next.

Pyloric cuff

A second method is to measure how drinking is affected by preventing water emptying from the stomach into the rest of the gut by tightening a cuff around the pylorus (figure 6.1). This effectively separates oropharyngeal and gastric ('orogastric') factors from intestinal, hepatic–portal and systemic factors. It is found that rats with a pyloric

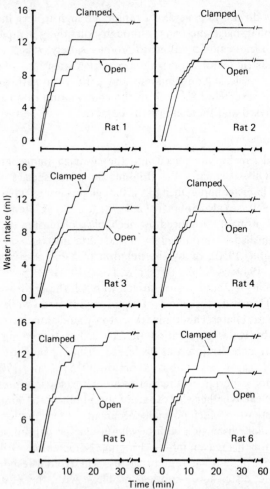

Fig. 6.8. Drinking patterns following deprivation for six rats with normal open stomachs or with the pyloric noose in the clamped state to prevent gastric emptying. Because the rats overdrank with the pylorus clamped closed, gastric distension cannot be sufficient normally to terminate drinking. (From Hall, 1973.)

cuff closed actually drink more than when the cuff (or noose) is open (Hall, 1973; Blass & Hall, 1976; Hall & Blass, 1977). This is illustrated in figure 6.8. This suggests, although further experiments are needed (Deutsch, 1979), that gastric distension, even in combination with oropharyngeal factors, does not provide a sufficient explanation for the termination of normal drinking in the rat. Some post-gastric factor (such as intestinal stimulation by water, or systemic dilution) must also be involved.

Relief of gastric distension

A third method for investigating the role of gastric factors in the termination of drinking is to allow drinking to terminate normally, and then immediately to relieve gastric distension by allowing the water which has accumulated in the stomach to drain from a gastric cannula. In recent experiments in the monkey, we have found that drinking starts again just after the gastric cannula is opened (see figure 6.9.) (Maddison, Rolls, Rolls & Wood, 1980). This provides quite convincing evidence that gastric distension is a factor involved in the termination of

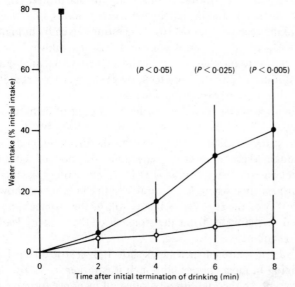

Fig. 6.9. Means and S.E. for water intake in the monkey following initial drinking termination (open circles, normal rehydration; closed circles, stomach drained at the time drinking terminated), and gastric drainage (closed square). Significant differences between conditions (one-tailed *t*-test) are shown. More drinking occurs if distension is relieved by drainage and therefore gastric distension must contribute to satiety. (From Maddison, Rolls, Rolls & Wood, 1980.)

drinking in the monkey. Although the experiment does not prove that distension itself must be the crucial stimulus, this is quite likely because of the short latency (often seconds) to the re-initiation of drinking. In this short time it is unlikely that, for example, altered duodenal perfusion because of a reduced gastric emptying rate would have time to become effective and influence drinking. This point could be settled with a pyloric cuff experiment which the reader might like to design. One strong feature of this type of evidence is that it cannot be argued that the gastric stimulus is abnormal, because the monkey is allowed to terminate drinking himself normally, and then it is demonstrated that removal of one factor, that of gastric distension, leads to the re-initiation of drinking. This proves that the factor (in this case probably gastric distension) was at least part of the mechanism inhibiting further drinking. It is of interest that, if there was a delay after drinking had terminated before the gastric cannula was opened, then the re-initiation of drinking was delayed for several minutes. This implies that after drinking has terminated, other factors (perhaps, for example, the presence of water in the intestine, or the on-going absorption of fluid from the gut) soon become effective in inhibiting further drinking.

The view that, at least in the primate, gastric distension is a factor in terminating drinking receives support from the ratings given by humans of their subjective sensations when they drink to satiety following 24 h water deprivation. It is found that the rating (on a 10-cm visual analogue scale) of 'How full does your stomach feel now?' changes very rapidly to indicate that the subjects experience feelings of a very full stomach within 2.5–5 min after access to water, as the main bout of drinking is terminating (figure 1.2; B. J. Rolls, Wood, Rolls *et al.*, 1980). The subjects even gave their reason for terminating the drinking as a full stomach. Although these subjective sensations do not of course necessarily reflect gastric distension, it is at least interesting that these sensations occur at a time when gastric drainage in the monkey shows that most (80–90%) of the ingested water is still in the stomach, and when withdrawal of the water from the stomach of the monkey leads to the rapid re-instatement of drinking.

In summary, gastric loading experiments show that gastric distension can inhibit drinking in most species (although it is rather ineffective in the dog). Experiments in the rat with a pyloric cuff or noose to prevent gastric drainage into the intestine indicate that gastric distension, even in combination with oropharyngeal factors, is not sufficient to account for the termination of drinking. Evidence, though, that gastric distension does contribute to the termination of drinking comes from experiments (in the monkey) in which the opening of a gastric fistula immediately when drinking has terminated leads to the re-instatement of drinking. Ratings by humans of stomach fullness when drinking terminates are consistent with the conclusion that (at least in the primate, if not the dog) gastric distension is an important factor in terminating drinking.

Intestinal and hepatic–portal factors in the termination of drinking

Given that there are osmoreceptors in the duodenum (Hunt, 1956; Hunt & Stubbs, 1975), and that some water reaches the duodenum and the rest of the intestine soon after the initiation of drinking, the intestine is one site where the effects of ingested water could be sensed, and could contribute to the termination of drinking. Similarly, fluid absorbed from the gut is carried to the liver in the hepatic–portal system, and sensors in this system could contribute to satiety. Ways in which the roles of these potential signals can be assessed are described next.

First, water can be prevented from reaching the duodenum and the

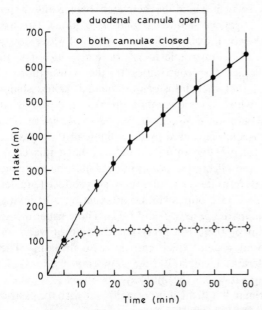

Fig. 6.10. Drinking in five monkeys after 22.5 h water deprivation is much greater, and is not terminated normally, when a duodenal cannula is open (closed circle) so that water does not reach the intestine, compared with normal drinking, i.e. with both cannulae closed (open circles). Thus, water must reach the duodenum for normal satiety to occur. (Mean cumulative intake ±S.E.). (From Maddison *et al.*, 1980).

rest of the intestine by opening a cannula inserted into the duodenum close to the pylorus (see figure 5.2). Experiments in the monkey show that, when ingested water drains in this way from the initial part of the duodenum, drinking of a great amount of water occurs, and drinking is clearly not terminated normally (figure 6.10; Maddison *et al.*, 1980). This shows that some factor at or beyond the level of the duodenum is necessary for the normal termination of drinking.

One way in which this factor acts is probably by controlling gastric emptying, for in the duodenal sham-drinking monkey (figure 6.10), water drains freely from the duodenal cannula and does not accumulate in the stomach (Maddison *et al.*, 1980). Because of the failure of this enterogastric control to operate, there is no gastric distension to contribute, in the way shown above, to the termination of drinking. In fact, by controlling gastric emptying, and thus allowing gastric distension to be produced by drinking, intestinal factors must play an important role in the termination of drinking.

A second way in which water reaching the intestine could contribute to satiety is by activating presystemic receptors in the intestine or hepatic–portal system (which would be activated before the water reached the 'systemic', or general, blood vascular circulation). This has been tested in the monkey by infusing different volumes of water slowly into the intestine (at rates comparable to gastric emptying rates) to determine directly whether this contributes to the termination of drinking. It was found that such infusions attenuated gastric sham-drinking (i.e. drinking with a gastric cannula open) in a dose-related manner (figure 6.11; Maddison *et al.*, 1980; Wood *et al.*, 1980). The effect was not due to discomfort caused by the volume of the infusate, in that even large isotonic saline infusions did not have this effect (figure 6.11). Further, the effect was not produced just by systemic dilution, in that the same infusate given intravenously produced greater systemic dilution, but produced only a similar attenuation of drinking (Wood *et al.*, 1980) (compare figures 6.5 and 6.11). These experiments indicate that in the monkey there is a presystemic action of water on the gut or hepatic–portal system which can decrease drinking. The experiments also underline incidentally the importance of gastric distension in producing satiety in the monkey, for instead of intake being limited at approximately 130 ml as it normally was with the gastric cannula closed (figure 5.3), with the gastric cannula open intake continued to more than 500 ml (figure 6.11) even though oropharyngeal stimulation was being provided and water was present in the gut.

The possibility that hepatic–portal mechanisms contribute to the termination of drinking has received little investigation, but Kozłowski

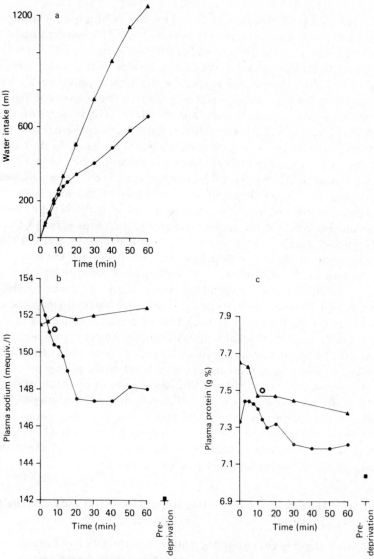

Fig. 6.11. The effect of 100 ml intestinal infusions of water (circles) or isotonic saline (triangles) on (*a*) drinking, (*b*) plasma sodium concentration, and (*c*) plasma protein concentration, in the monkey. The infusions were given at 5 ml/min starting at time 0, the time when sham-drinking, with drainage from the stomach, was allowed to start. The infusions of water (but not of saline) reduced the drinking rate, and the reduction was similar to that produced by intravenous infusions of water, even though the latter produced greater systemic dilution (see text; compare with figure 6.5 and also with figures 6.2 and 6.3). This suggests that there is a presystemic, intestinal or hepatic–portal contribution to satiety. (After Wood *et al.*, 1980.)

& Drzewiecki (1973) did report that, in the dog, hepatic–portal infusions of water increase the osmotic threshold to drink. The osmotic threshold was measured by the amount of hypertonic solution which had to be infused to initiate drinking. The finding implies that, if water is being absorbed from the gut and is entering the hepatic–portal circulation, then the water-deprived dog becomes less sensitive to systemic signals indicating cellular dehydration, and therefore tends to stop drinking earlier. Clearly further investigation is needed, but consistent with this possibility is the observation that radioactive water given via the mouth enters the hepatic–portal circulation within 2–3 min in the dog (Kozłowski & Drzewiecki, 1973). Further, there is electrophysiological evidence for the existence of osmoreceptors in the hepatic–portal system, in that osmotically active infusions into the hepatic–portal circulation influence the activity of fibres in the vagus nerve (Adachi, Niijima & Jacobs, 1976) and in the hypothalamus (Schmitt, 1973). Also, osmotically active hepatic–portal infusions can influence the release of antidiuretic hormone (Chwalbinska-Moneta, 1979).

In conclusion, the overdrinking which occurs in duodenal sham drinking in the monkey shows that the passage of water into the intestine is necessary for the normal termination of drinking. One effect of the passage of water into the intestine is to limit the rate of gastric emptying into the duodenum, so that gastric distension occurs, and contributes to the termination of drinking. Another effect of the passage of water into the intestine is to attenuate drinking, as shown by the decrease in the gastric sham-drinking rate produced by intestinal infusions. This effect is not due only to systemic effects occurring after absorption of the water which enters the intestine, as shown by intravenous infusion experiments. The effect could be mediated by a direct effect on the intestine itself, or via hepatic–portal mechanisms.

The termination of drinking initiated by cellular and extracellular stimuli

So far, we have been considering mainly the termination of water intake initiated in response to water deprivation. It has been shown, for example, that after water deprivation in the rat and dog, both cellular and extracellular fluids are depleted, and that the cellular fluid depletion accounts for 60–70% of the drinking and the extracellular fluid depletion for 20–30% of the drinking, on the basis of deficit replacement experiments (Chapter 4). But what happens if there is overexpansion of the extracellular fluid compartment? Does this turn off drinking in response to cellular stimuli of thirst? Or, if the more dangerous situation of overdilution occurs, with the possibility of water intoxication, nausea and confusion, does this turn off drinking induced by depletion of the extracellular fluid? These questions can be answered by the following types of investigation.

Cellular stimuli

First, it is known that if cellular dehydration is produced by injections of hypertonic saline, bilaterally nephrectomized rats drink just enough water to dilute the salt load to isotonicity (see Chapter 4 and figure 4.2; Fitzsimons, 1961a), i.e. to remove the cellular dehydration. This shows that cellular rehydration turns off drinking induced by cellular dehydration, as expected. But the nephrectomy prevents excretion of any of the salt load and the salt cannot enter the cells, so that at the termination of drinking there is expansion of the extracellular fluid volume. Yet this expansion of the extracellular fluid volume is clearly ineffective in terminating drinking in response to the cellular stimulus of thirst, as drinking continues until the salt load is diluted. Thus, expansion of the extracellular fluid volume does not inhibit drinking induced by cellular dehydration. The same conclusion follows from experiments in which overexpansion of the extracellular fluid volume was produced by intravenous infusions of isotonic saline in the dog. The over-expansion did not inhibit drinking induced by water deprivation significantly more than simple correction of the extracellular fluid deficit did (Ramsay, Rolls & Wood, 1977a; compare the effects of infusions of saline which corrected the deficit (D/2) to those which overexpanded the extracellular compartment (D) in figure 4.22).

Extracellular stimuli

Second, the termination of drinking caused by an extracellular stimulus is considered. In this case, what is needed to replace the deficit is isotonic saline, and if rats are given isotonic saline to drink, or hypertonic saline as well as water, they drink sufficient saline to replace the deficit (Stricker & Jalowiec, 1970). Thus, expansion of the depleted extracellular fluid volume, and removal of the stimulus, is one straightforward way in which extracellular drinking can be terminated. If only water is available to drink when the thirst stimulus is extracellular, then overdilution of the body fluids and cellular overhydration occurs, and this terminates drinking rapidly. In man, water intoxication with nausea and vomiting would occur (see Chapter 9) if this overdilution were not effective in terminating drinking. When drinking induced by an extracellular stimulus does terminate, there is still likely to be an extracellular fluid deficit, as all the absorbed water distributes itself throughout both the cellular and extracellular space. The effectiveness of the cellular system is emphasized by the fact that water-deprivation-induced drinking in the dog is terminated by intracarotid infusions which bring the osmolality sensed by the brain to the predeprivation level, even though the extracellular fluid depletion produced by the water deprivation is not affected (Ramsay, Rolls & Wood, 1977a) (figure 4.22, Chapter 4). One way in which animals can prevent overdilution when drinking water to restore an extracellular fluid deficit is by eating, as this will provide the solutes necessary, with water, to replace the extracellular fluids.

It can be concluded that drinking in response to cellular stimuli is finally terminated when the cellular dehydration is restored by absorbed water, and that overexpansion of the extracellular fluid volume is relatively ineffective in inhibiting cellular thirst. Drinking produced by extracellular fluid deficits is terminated by replacement of the deficits, or will be terminated by dilution even if the extracellular deficit has not been replaced.

Conclusions

In most species, except perhaps in very slow drinkers, systemic changes in osmolality and volume do not occur rapidly enough to account for the termination of drinking. Oropharyngeal factors, gastric distension, and intestinal (and/or hepatic–portal) factors all contribute to the normal termination of drinking. The relative importance of these factors varies among different species, with gastric distension being relatively unimportant in the dog, but necessary in the monkey. The operation of these, mainly pre-absorptive, factors implies that metering of the amount of water ingested occurs in that ingestion is often close to that required to restore the deficit. Although these mainly pre-absorptive factors are important in the early termination of drinking, absorption of water from the gut starts to have measurable systemic effects within a very few minutes of the onset of drinking, and these soon become large enough (as assessed by infusion experiments) to contribute to the inhibition of drinking, as other, for example oropharyngeal and gastric, factors decline in importance. Thus the termination of drinking is produced by the operation of a number of factors, which normally occur in a particular sequence and probably alter in importance as time since the onset of drinking progresses. The sequence in which the factors operate is probably important for their normal operation, in that for example the drinking of water is maximally satiating only when it occurs in conjunction with cellular overhydration (Blass & Hall, 1974).

Although the termination of drinking is normally analysed when food is not available, so that the operation of particular variables can be studied, food may be available in the natural environment. Under conditions of extracellular fluid depletion (which will for example be produced, together with cellular dehydration, by water deprivation) eating will provide solutes to be absorbed with ingested water, enabling the required isotonic replacement of the lost fluid to occur. Although animals must have a system which terminates drinking before over-dilution occurs, if water intake does exceed water need, then of course the additional water can be excreted in a hypotonic urine. Most animals do not normally 'overdrink' in this way, but when an abundance of palatable liquid refreshment is available, 'overdrinking' and the production of a dilute urine occurs commonly in man and may occur in other animals (see Chapter 10).

7   The neural control of drinking

Introduction

All the different signals which can contribute to the initiation or termination of drinking must be integrated by the brain, and a co-ordinated sequence of behaviour leading to appropriate drinking must be produced. Some of the signals are sensed peripherally, for example gastric distension, and some are sensed centrally, in the brain. Evidence that for example cellular dehydration is sensed centrally is provided by the observation that intracarotid infusions of hypertonic saline or sucrose induce drinking in the dog. Moreover, this central 'osmoreception' is sufficient to account for drinking in response to osmotic thirst stimuli, in that drinking to systemically (intravenously) administered hypertonic saline was abolished by intracarotid infusions of water just sufficient to remove the central osmotic stimulus (Wood, Rolls & Ramsay, 1977). Investigations, and methods of investigation, into which different parts of the brain are involved in sensing directly or responding indirectly to the different signals which control drinking behaviour, and in producing appropriate drinking behaviour, are described in this chapter. The brain mechanisms which alter drinking behaviour by learning on the basis of previous experience are also important to our understanding of the control of drinking in the natural environment, and are considered. Figure 7.1 shows some of the brain regions which will be discussed. Basic approaches to brain function and behaviour are described elsewhere (e.g. Carlson, 1977).

The location of central osmoreceptors for thirst

Andersson (1953), in a pioneering study, showed that injections of hypertonic saline into the medial anterior hypothalamus of goats elicited drinking. Andersson & McCann (1955) went on to show that drinking was elicited by relatively large (3–10 μl), and rather highly hypertonic (2–3%, compared to the isotonic value of 0.9%), saline injections into the hypothalamus near the paraventricular nucleus and by electrical stimulation in the same region. However, the actual site of action in the goat is still not clear, for infusions into the nearby anterior ventral part of the third ventricle, from which the infusion could spread, were very effective in eliciting drinking (Andersson, Dallman & Olsson, 1969).

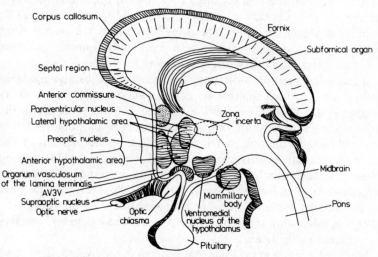

Fig. 7.1. Sagittal three-dimensional representation of the brain to illustrate some brain regions implicated in the control of drinking. AV3V: antero-ventral region of the third ventricle.

The evidence that an osmoreceptor system for thirst is located just anterior to this region, in the preoptic area, in the rat and rabbit has been described above, in Chapter 4, pp. 36–9. Briefly, the evidence came from studies showing that (1) small, mildly hypertonic injections into the preoptic area elicited drinking, (2) the same injections did not elicit eating, (3) small injections of water suppressed drinking induced by systemic cellular dehydration, and (4) lesions in the head of the preoptic nucleus abolished drinking elicited by systemic stimuli, but had little effect on drinking elicited by extracellular stimuli (and were therefore relatively specific) (Blass & Epstein, 1971; Peck & Novin, 1971; Blass, 1974). However, also in these studies the exact location of the central osmosensitive zone for thirst is not clear, for drinking was not elicited from all injection sites in the preoptic area and was elicited from some other sites throughout the hypothalamus, distributed from the anterior commissure anteriorly to the zona incerta caudally and from the midline to the internal capsule laterally (Peck & Blass, 1975). Walsh & Grossman have also elicited drinking in rats with hypertonic injections as far caudal as the zona incerta (personal communication) and have found that lesions here can impair drinking, particularly in response to cellular dehydration (Walsh & Grossman, 1973, 1976, 1977, 1978). However, lesions in the zona incerta produce subtle motor or sensory-motor deficits which may impair drinking by reducing the volume of water obtained per lap, rather than by impairing the ability to respond to water-deficit signals (Evered & Mogenson, 1976, 1977; Brimley & Mogenson, 1979). The tissue surrounding the anteroventral part of the third ventricle has also been implicated in osmoreception for thirst in the rat, for periventricular lesions here depress responses to systemic cellular dehydration (Buggy & Johnson, 1977b), and injections of hypertonic sodium chloride or sucrose solutions here elicit drinking (Buggy *et al.*, 1979). Confirmation that the osmoreceptors are in the region of the anterior third ventricle comes from lesion studies in larger animals such as the dog and sheep. In these species, lesions of the circumventricular organ, the organum vasculosum of the lamina terminalis, significantly reduce the drinking in response to plasma hypertonicity (Thrasher, Simpson & Ramsay, 1980; McKinley, Denton, Graham *et al.*, 1980). More work is needed to clarify the role of the OVLT in thirst since the daily fluid balance of these dogs and sheep is not affected by the lesions.

Electrophysiological evidence that neurons in the hypothalamus and preoptic area are influenced by, and may even respond directly to, osmotic stimuli is described in Chapter 4.

Thus, there is considerable evidence that there are osmoreceptors for thirst in periventricular tissue near the anteroventral part of the third ventricle, for example in the preoptic area or anterior hypothalamus, but perhaps extending as far posterior as the zona incerta. How restricted this osmoreceptor area is or exactly which nervous tissue performs this function is a topic for future investigation. It is likely, however, that the osmoreceptors for thirst are different from those located in or near the supraoptic nuclei for ADH release, for in the studies of Andersson and of Peck & Blass, cited above, injections into the region of the supraoptic nuclei did not induce drinking (see also Andersson, 1978).

Neural systems for extracellular thirst

The evidence that angiotensin in the systemic circulation acts centrally, and that the site of action is in the subfornical organ, or the organum vasculosum of the lamina terminalis in the region of the anteroventral part of the third ventricle, or possibly the preoptic area, is described in Chapter 4, pp. 48–52. In order to understand how the drinking behaviour is finally produced, it is necessary to identify the output systems of these circumventricular organs. Efferent projections of the subfornical organ are now starting to be identified (Miselis, Shapiro & Hand, 1979). There are projections to the nucleus medianus of the medial preoptic area, to the organum vasculosum of the lamina terminalis, and to the supraoptic and paraventricular nuclei, the nucleus circularis, the perifornical nucleus, and the periventricular area of the hypothalamus (R. Miselis, personal communication, 1980). It is possible, therefore, that these connections provide part of the basis by which angiotensin induces drinking, the release of antidiuretic hormone, and pressor responses. It should be recalled that after preoptic–hypothalamic lesions in the periventricular tissue in the region of the AV3V, there are persistent drinking deficits in response to angiotensin (Buggy & Johnson, 1977a, b). The same lesions produce a period of adipsia (lack of drinking) after the lesions, and also persistent drinking deficits after systemic hypertonic saline, reduced drinking in response to water deprivation, and impaired antidiuretic responses (Buggy & Johnson, 1977a, b; Johnson & Buggy, 1978). Also recall that injections of hypertonic solutions, as well as of angiotensin, into the OVLT/AV3V region induce drinking (Buggy *et al.*, 1979). All this evidence suggests that in the preoptic–hypothalamic region, close to the ventricles, there are neuronal circuits concerned with responses to cellular and to at least some extracellular thirst stimuli. Drinking induced by polyethylene glycol was not impaired by the AV3V lesions (Buggy & Johnson, 1977b). Now, drinking in response to this extracellular thirst stimulus is not abolished by nephrectomy, so it does not depend on angiotensin (Stricker, 1973). Much less is known about the neural circuitry for drinking mediated by extracellular factors other than angiotensin. For drinking induced by some extracellular thirst stimuli, it is possible that distension receptors in the low-pressure circulation influence hypo-thalamic neuronal activity via the vagus nerve. Clearly, there is considerable work waiting to be done on exactly how both extracellular and cellular stimuli of drinking influence the brain.

The lateral hypothalamus and preoptic area

Since Teitelbaum & Stellar (1954) reported that small bilateral lesions of the lateral hypothalamus in the rat produced adipsia (lack of drinking) as well as aphagia (lack of eating), there has been great interest in the role of the hypothalamus in drinking. The adipsia and aphagia were so severe that the rats died unless they were maintained by intra-gastric injections of water and food, but eventually some recovery usually occurred. This gradual recovery proceeds over days or weeks through stages in which wet, palatable food is accepted and eventually the animals manage to maintain themselves on dry food and water (Teitelbaum & Epstein, 1962; Epstein, 1971). However, the normal controls of drinking do not recover in these rats. For example, when deprived of food, or when water but not food is available after a period of water deprivation, the rats drink almost nothing. Instead, the rats drink when food is available, taking a draught of water with every one or two bites of food in order to enable them to chew and swallow the food (Teitelbaum & Epstein, 1962). Part of the reason why this 'prandial' (i.e. feeding-associated) drinking does occur may be inadequate salivary secretion by the lesioned rat (Kissileff & Epstein, 1969). In addition to failing to drink normally after water deprivation, the rats with bilateral lateral hypothalamic damage also fail to drink in response to a cellular thirst stimulus (hypertonic saline), to an extra-cellular thirst stimulus (polyethylene glycol) or to hyperthermia (Epstein & Teitelbaum, 1964; Stricker & Wolf, 1967; see Epstein, 1971).

Electrical stimulation of the lateral hypothalamus can produce drinking (Andersson & McCann, 1955; Greer, 1955; see E. T. Rolls, 1979) as well as eating. There is some evidence that the sites from which drinking as opposed to feeding are elicited can be dissociated (Olds, Allan & Briese, 1971; Huang & Mogenson, 1972), but also evidence that stimulus-bound drinking can 'switch' to stimulus-bound eating if water is replaced with food, and vice versa (Valenstein, Cox & Kakolewski, 1970). Because of this 'behavioural plasticity', it is difficult to obtain strong evidence from stimulus-bound motivational behaviour that a specific substrate for drinking (separate, for example, from one concerned with feeding) is present in the lateral hypothalamus. Indeed, the observation that the simple application of continuous pressure from a paper clip on a rat's tail can elicit stimulus-bound feeding or drinking in an environment in which food or water is available (Antelman, Rowland & Fisher, 1976) also suggests that stimulus-bound feeding and drinking elicited by lateral hypothalamic stimulation should not be overinterpreted. It would be difficult to prove that they do not arise from some relatively non-specific activation. Clearly, this does not disprove a role for the lateral hypothalamus in feeding and drinking, but just advises caution in interpreting feeding and drinking elicited by hypothalamic stimulation. In the context of stimulation of the lateral hypothalamus, it may be noted that injections of hypertonic saline into the lateral hypothalamus elicit feeding and drinking, so that this appears to be due to a relatively non-selective activation (Blass & Epstein, 1971). In contrast, hypertonic injections into the preoptic area elicit drinking, but not eating (Blass & Epstein, 1971; Peck & Novin, 1971), a result that provides evidence for osmoreceptors for drinking in this area.

There are many fibre systems passing through the lateral hypothalamus. The question therefore arises of the extent to which the thirst deficits produced by lateral hypothalamic lesions arise because of destruction of the cell bodies in the lateral hypothalamus, or from damage to the fibre pathways which course through the lateral hypothalamus. There is evidence that damage to some of the monoamine-containing pathways, passing from their cell bodies of origin in the brainstem through the hypothalamus to the forebrain (figure 7.2), accounts partly for, but may not explain fully, the effects of lateral hypothalamic damage on drinking and feeding. In particular, damage to the dopamine pathways can produce aphagia and adipsia. The nigrostriatal dopaminergic pathway projects from the A9 monoamine cell group in the substantia nigra, pars compacta, through the far lateral part of the hypothalamus to the striatum, in particular to the caudate nucleus and putamen (figure 7.2). The mesolimbic and meso-cortical dopaminergic pathways project from the A10 monoamine cell group just medial to the substantia nigra, near the interpeduncular nucleus in the 'limbic midbrain area', through the lateral hypothalamus, to the nucleus accumbens (see figure 7.2), and frontal and cingulate cortices, respectively (see Ungerstedt, 1971; Livett, 1973; Lindvall & Björklund, 1978). These dopaminergic pathways can be selectively damaged with the neurotoxin 6-hydroxydopamine, and if appropriate pretreatment is given there is little damage to the nearby noradrenergic pathways. Destruction of the dopaminergic pathways with 6-hydroxydopamine injections into the cerebral ventricles, into the substantia nigra, or on the course of the nigrostriatal bundle, leads to adipsia and aphagia, similar in some respects to that seen in the lateral hypothalamic syndrome (Ungerstedt, 1971; Marshall, Turner & Teitelbaum, 1971; Marshall, Richardson & Teitelbaum, 1974). The thirst deficits include reduced drinking during food deprivation and in response to cellular (hypertonic saline) and extracellular (polyethylene glycol) thirst stimuli (Marshall, Richardson & Teitelbaum, 1974; Stricker & Zigmond, 1974, 1976). However, damage to the dopaminergic pathways does not account fully for the lateral hypothalamic syndrome, as the following evidence shows. First, 'recovered' rats with bilateral lateral hypothalamic lesions show 'finickiness', in that they show exaggerated aversion to diets or solutions containing unpleasant-tasting substances such as quinine. The catecholamine-depleted rats show little of this finickiness (Smith, 1973; Marshall, Richardson & Teitelbaum, 1974). This suggests that the lateral hypothalamus is in some way particularly involved in behavioural reactions to the taste of food (see below). Second, even extreme

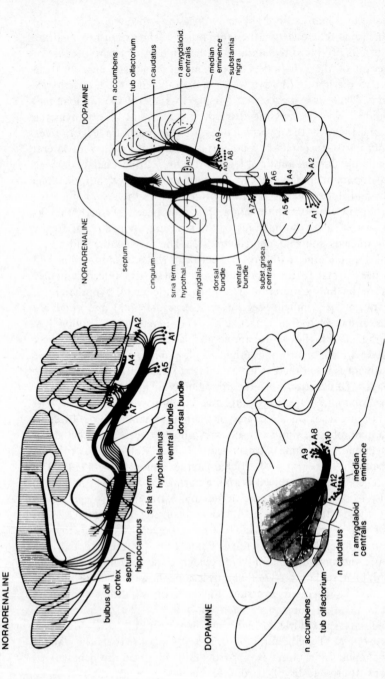

Fig. 7.2. Some of the noradrenaline-containing and dopamine-containing pathways in the rat brain. There are dopaminergic and noradrenergic pathways which ascend through or close to the hypothalamus. (After Ungerstedt, 1971; from Livett, 1973.)

destruction of dopamine pathways produces less severe long-term deficits in regulation than lateral hypothalamic lesions which produce only a relatively modest depletion of forebrain dopamine (Zigmond & Stricker, 1973).

In a different sort of approach to the function of the dopamine pathways in feeding and drinking, we made recordings from single neurons in the substantia nigra in the monkey during feeding and drinking (Mora, Mogenson & Rolls, 1977). It was found that some neurons were active during feeding and drinking, but that their activity was related to movements, such as mouth and arm movements, and not to the animal's motivational state. Although these were not necessarily dopaminergic neurons, the dopaminergic neurons have dendrites in the region of the non-dopaminergic neurons in the pars reticulata of the substantia nigra, through which they can probably be influenced. Further, consistent evidence on the function of the nigrostriatal dopamine neurons comes from recordings made from single neurons in the region to which the nigrostriatal system projects, namely, the caudate nucleus and putamen. Neurons in the caudate nucleus and putamen have activity related to movements, or the preparation for and enabling of movements, rather than to motivational state (E. T. Rolls, Thorpe *et al.*, 1979; E. T. Rolls, 1981a). Thus, neurophysiological evidence indicates that the nigrostriatal projection may be part of a system concerned with preparation for movement and the control of movement, rather than with the motivational control of feeding or drinking. Damage to the nigrostriatal system may thus, on this basis, interfere with feeding and drinking by impairing the ability of the animal to initiate feeding and drinking responses (E. T. Rolls, Perrett *et al.*, 1979; E. T. Rolls, 1980, 1981a, b).

Although this type of impairment may account partly for the effects of lateral hypothalamic lesions, as shown above, it does not fully account for the effect of lateral hypothalamic lesions on drinking. The problem with the evidence from lateral hypothalamic lesions is that, in the presence of damage to several different fibre and cell systems, some of which have now been shown to be concerned with the initiation of behavioural responses, it is very difficult to draw conclusions about whether there is a hypothalamic system specifically concerned with drinking, and if so, about how it functions. The method of recording the activity of single neurons is then particularly appropriate for analysing the functions of the hypothalamus, as it consists of intermingled populations of neurons with different types of response, all interspersed between different ascending and descending fibre pathways (see E. T. Rolls, 1980, 1981a, b).

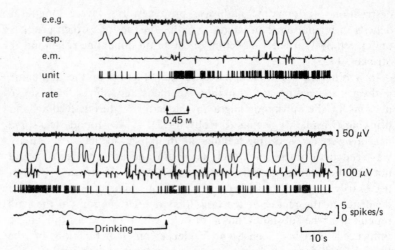

Fig. 7.3. When hypertonic saline (0.45 M) was infused via a catheter into the carotid artery, the discharge rate (rate) of the neuron (unit) in the hypothalamus increased. The firing rate of this same neuron was decreased during drinking (lower record). The other recordings are of electroencephalogram (e.e.g.), respiration (resp.) and eye movements (e.m.). (After Vincent, Arnauld & Bioulac, 1972.)

Vincent, Arnauld & Bioulac (1972) have described neurons in the lateral and dorsal hypothalamus of the monkey, which respond to the intracarotid injection of hypertonic saline (figure 7.3). If not necessarily osmoreceptors themselves, the neurons were at least influenced by osmotic stimuli. Although these neurons could be part of a system for controlling neuronal activity in the supraoptic nucleus, and thus the release of antidiuretic hormone (see Arnauld, Dufy & Vincent, 1975), the authors suggested that, alternatively, the neurons could be involved in drinking induced by cellular dehydration (Vincent, Arnauld & Bioulac, 1972). Interestingly, the same neurons responded in the opposite direction (by decreasing their firing rate) when the monkey drank water. Consistent with these findings, Emmers (1973) has described neurons in the cat hypothalamus which were influenced both by intracarotid infusions of 3% sodium chloride solution or distilled water, and by electrical stimulation of gustatory and splanchnic relay regions of the thalamus. This response could be related to the 'feedforward' (or anticipatory) effect which water on the tongue has in turning off the antidiuretic response to systemic signals of cellular dehydration (see also Nicolaïdis, 1969), or equally perhaps to the partial satiety produced by oropharyngeal stimulation by water (see Chapter 6). Further evidence that sensory inputs related to drinking

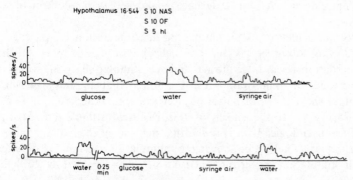

Fig. 7.4. Firing rate of a hypothalamic neuron which responded by increasing its firing rate when the hungry and thirsty monkey tasted and drank water, but showed little response when 5% glucose (or isotonic saline) was tasted and drunk, or when air was blown on the mouth from a syringe (this was aversive). The neuron fired most rapidly while the water was in the mouth. Time scale: the stimuli were presented for approximately 20 s. The same neuron was activated trans-synaptically during self-stimulation of the nucleus accumbens (NAS), orbitofrontal cortex (OF), and lateral hypothalamus (hl) with the latencies (in ms) shown.

(and others related to feeding) reach the hypothalamus comes from other neurophysiological studies by E. T. Rolls and his colleagues (E. T. Rolls, 1974, 1975, 1980, 1981a, b; E. T. Rolls, Sanghera & Roper-Hall, 1979; E. T. Rolls, Burton & Mora, 1980). In these studies it was found that some hypothalamic neurons in the monkey were influenced by the taste of water in the mouth (figure 7.4), and that others responded before drinking, when the monkey was shown a syringe from which he was given water to drink. While some neurons were activated by the sight of water or the sight of food, others were activated only by the sight of food (E. T. Rolls, Sanghera & Roper-Hall, 1979). There appear to be overlapping populations of neurons, and it could well be that in future studies neurons which respond primarily to the sight of water will be found. Those which respond primarily to the sight of food respond with relatively short latencies (150–200 ms), compared with the feeding responses with latencies of 300 ms made by the monkeys (E. T. Rolls, Sanghera & Roper-Hall, 1979), and only respond if the monkey is hungry (Burton, Rolls & Mora, 1976). The responses of these neurons thus precede and predict the responses made by the hungry monkey to food (E. T. Rolls, 1981b; E. T. Rolls & Rolls, 1981). These findings that hypothalamic neurons are activated by osmotic stimuli, and by sensory inputs related to drinking such as the taste and sight of water, suggest that the hypothalamus is involved in integrating internal and external signals used in the control of water intake.

Some evidence on the nature of this integration comes from the following observations on the neurophysiological basis of brain-stimulation reward, or intracranial self-stimulation. With stimulation electrodes in some brain sites, it is found that animals will work to obtain short trains of pulses of electrical stimulation (see E. T. Rolls, 1975, 1979). If the effects of this electrical stimulation are traced neurophysiologically, it is found that neurons in the hypothalamus are particularly likely to be activated by stimulation at a number of different self-stimulation sites if these neurons are also activated by the sight or taste of food or water (E. T. Rolls, Burton & Mora, 1980). This convergence of effects onto individual hypothalamic neurons is interesting, for sensory inputs, such as the sight and taste of water or food, are present during sham drinking or feeding and are the stimuli which normally maintain or reward or provide the reinforcement or incentive for normal drinking or feeding (see Chapter 5, and E. T. Rolls, 1975). Thus this convergence of the effects of brain-stimulation reward and food or water reward onto hypothalamic neurons suggests that their function is related to the incentive value which the sight and taste of food or water have for the hungry or thirsty animal. For the feeding-related neurons, there is further evidence. These neurons only respond to the sight or taste of food if the monkey is hungry (Burton, Rolls & Mora, 1976; E. T. Rolls, Burton & Mora, 1980; E. T. Rolls & Rolls, 1981), that is, when the food is rewarding. Further, self-stimulation can be obtained in the region of these neurons, and is attenuated when the monkey is satiated, as opposed to hungry (E. T. Rolls, Burton & Mora, 1980). All this evidence is consistent with the hypothesis that these hypothalamic neurons are involved in the integration of sensory inputs produced by the sight and taste of food with internal signals of the motivational state of the animal. The result of this is that a signal which could control autonomic, endocrine and feeding responses on the basis of whether food is rewarding is produced (E. T. Rolls, 1975, 1976, 1978, 1979; E. T. Rolls, Burton & Mora, 1980; E. T. Rolls & Rolls, 1981). Although there has been less investigation of hypothalamic neuro-physiology related to the ingestion of water, and more is needed, it is clear that some hypothalamic neurons are activated by the taste and/or the sight of water, and there is some evidence for an interaction with the thirst of the animal (see above). There is also evidence, from the rat, that electrical stimulation of the hypothalamus can be equivalent to water for a thirsty animal. The evidence comes from an experiment by Gallistel & Beagley (1971), in which rats chose to self-stimulate at one hypothalamic site if thirsty, but at a different site if hungry. This shows that the effects of brain-stimulation reward in the lateral hypothalamus can be specific, mimicking either water, or food, at different sites (see E. T. Rolls, 1975, 1979).

Thus, for thirst as well as for hunger, there may be a system of hypothalamic neurons for integrating the effects of sensory stimulation produced by the taste and even the sight of water with internal signals reflecting the thirst of the animal. The thirst signals could be derived from the osmoreceptors that are thought to be located in the peri-ventricular preoptic region and hypothalamus. The output of this system could be involved in the responses (such as drinking) produced

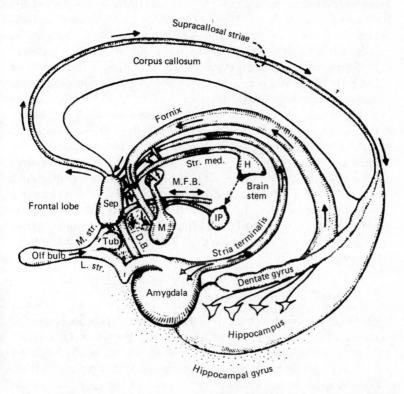

Fig. 7.5. Some limbic structures and connections are shown. L.Str., lateral olfactory tract; M.Str., medial olfactory tract; Tub, olfactory tubercle; Sep, septal area; D.B., diagonal band of Broca; M.F.B., medial forebrain bundle; IP, interpenduncular nucleus (in ventral tegmental area); H, habenular nucleus; Str.med., stria medullaris; M, mammillary body; At, anterior nucleus of the thalamus. The medial forebrain bundle traverses anteriorly the preoptic area and hypothalamus. (After a drawing by P. D. MacLean.)

by water when it is rewarding in the thirsty animal. A comparable system located near the supraoptic neurons may control the activity of the supraoptic neurons, and thus the release of antidiuretic hormone.

Thus the neurophysiological evidence suggests that hypothalamic neurons are involved in the regulation of water intake. The evidence obtained with this method emphasizes, however, that only a relatively small number of hypothalamic neurons have activity specifically related to feeding (13.6% in one sample of 764 neurons – see E. T. Rolls, Burton & Mora, 1976, 1980), and fewer are specifically related to drinking. Further, it is clear that the neurons with this type of response are widely distributed, and are found in the preoptic area, the lateral hypothalamus and the laterally adjacent substantia innominata (E. T. Rolls, Sanghera & Roper-Hall, 1979). This heterogeneity, and the presence of fibres of passage, makes study by the lesion method particularly difficult for the hypothalamus. But it is quite consistent with the neurophysiological evidence that injections in the region of the hypothalamus of kainic acid, which destroys cells but leaves fibres of passage relatively intact, produce feeding and drinking deficits (Stricker, Swerdloff & Zigmond, 1978; Grossman *et al.*, 1978; Brown & Grossman, 1980). In particular, lasting deficits in drinking responses to cellular dehydration (tested with hypertonic saline injections) and to extracellular fluid depletion (tested with polyethylene glycol) were obtained. All these findings suggest that further work on the integrative function of the hypothalamus in drinking, and on how the hypothalamus is influenced by sensory inputs related to drinking as well as thirst signals, and on the outputs of these hypothalamic systems will be helpful in understanding the organization of drinking behaviour (see also E. T. Rolls, 1980, 1981a, b).

Limbic and related structures

Limbic structures such as the amygdala and septal region are connected to the hypothalamus and preoptic area (see e.g. Hamilton, 1976; figure 7.5) and are thus in a position to influence mechanisms which control water intake. Some evidence that they can influence drinking follows.

Following up earlier evidence that lesions of the amygdala in the cat, dog and rat can decrease or increase drinking, B. J. Rolls & Rolls (1973) tested whether rats with amygdaloid lesions responded to cellular and extracellular thirst signals. Rats with bilateral lesions in the basolateral amygdala had no deficits in drinking in response to hypertonic saline, the cellular stimulus used, or to isoproterenol, which probably acts through extracellular thirst mechanisms. However, the lesioned

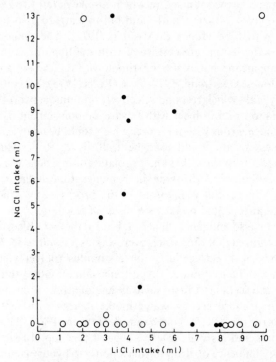

Fig. 7.6. Rats with basolateral amygdala lesions have a deficit in learned aversion. In a first part of the experiment (abscissa), rats drank 0.12 M lithium chloride. When retested later, the control non-lesioned, rats (open circles) drank very little similar-tasting sodium chloride (ordinate), having learned to avoid the solution as a consequence of the mild illness induced after the lithium chloride was ingested. However, rats with basolateral amygdala lesions (closed circles) drank the sodium chloride solution, failing to show normal learned aversion. (From B. J. Rolls & Rolls, 1973.)

rats were impaired when they had to learn to avoid ingesting a poisonous solution of lithium chloride. In the test of learned aversion that was used, the rats were given lithium chloride to drink. Because of the mild illness produced as a consequence of its ingestion, the normal, control, rats learned to avoid a similar-tasting solution of sodium chloride when it was given to drink 3 days later. However, the rats with basolateral amygdala lesions failed to learn normally to avoid the solution, the taste of which had been paired with subsequent nausea (figure 7.6; B. J. Rolls & Rolls, 1973). A similar impairment in learned taste aversions after basolateral amygdala lesions was found by Nachman & Ashe (1974). This learning deficit apparent in rats after amygdala lesions is comparable to the deficit produced by amygdala or anterior temporal lobe damage in the monkey, in which monkeys fail to avoid placing non-food objects in their mouths and fail to learn to associate a visual stimulus with food (Jones & Mishkin, 1972). The different effects produced by the lesions are consistent with the hypothesis that amygdala lesions impair the association of stimuli with reinforcers such as food or water (Jones & Mishkin, 1972; E. T. Rolls, 1975). Although neurons in the amygdala which respond selectively and uniquely on the basis of the association of an object with reward or punishment have not been found, some neurons with responses which could be at an early stage of this processing are found (Sanghera, Rolls & Roper-Hall, 1979). In conclusion, it appears that the basolateral amygdala is not involved in the regulation of drinking in response to cellular and extracellular thirst stimuli, but is involved in the processes by which previous experience affects fluid intake as a result of learning.

Septal lesions can also influence fluid intake. In the rat, Blass & Hanson (1970) have found that rats which overdrink as a result of septal lesions overrespond to an extracellular thirst stimulus (polyethylene glycol), but not to a cellular stimulus (hypertonic saline). Other thirst stimuli related to extracellular thirst, such as angiotensin, renin and caval ligation, also produce greater water intake in rats with septal lesions than in control rats (Blass, Nussbaum & Hanson, 1974). These results have led Blass and his colleagues to suggest that septal lesions remove an inhibitory influence on drinking in response to hypovolaemia. At present, the way in which lesions of this region produced increased water intake after extracellular thirst stimuli is not clearly understood. It can be noted here that there is little evidence at present for a satiety centre for thirst in the hypothalamus or median eminence region (see pp. 139–40.

Conclusions

There is evidence that cellular stimuli for thirst are sensed in or near periventricular tissue in the preoptic area and hypothalamus. The circumventricular organs, in particular the subfornical organ, and the organum vasculosum of the lamina terminalis in the anteroventral region of the third ventricle (and perhaps the preoptic region) appear to have receptors outside the blood–brain barrier for angiotensin in the systemic circulation. The subfornical organ has pathways to the AV3V region, to the preoptic area and to the supraoptic nucleus, and thus provides one route by which extracellular fluid deficits could influence drinking and the release of antidiuretic hormone. In the hypothalamus and preoptic area, there are neurons influenced by the oropharyngeal and even visual stimuli obtained when animals drink water. The same neurons can be influenced in the opposite direction by osmotic changes. These neurons thus appear to be involved in integrating the different signals utilized in drinking behaviour. Their responses could be involved in the responses produced by water in the thirsty animal, which include the maintenance of drinking by providing the reward for drinking (see Chapter 5). This is consistent with the finding that these hypothalamic neurons are activated by brain-stimulation reward, and that brain-stimulation reward, which at some sites depends on thirst, occurs in this region. The amygdala appears to be involved in the effects which learning has on fluid intake, and does not appear to be involved in responses to thirst stimuli.

8 Pharmacology of drinking

Introduction

A number of pharmacological agents stimulate drinking. Some may act directly on the brain, influencing neurotransmitter mechanisms which could be in the neuronal circuits involved in drinking. For example, when injected into some regions of the rat brain, cholinergic substances elicit drinking, and adrenergic substances such as noradrenaline (norepinephrine) elicit eating (Grossman, 1960, 1962a, b). This suggested that there might be a simple pharmacological coding of different types of ingestive behaviour. But it is certainly too simple to consider that cholinergic or noradrenergic pathways have any exclusive role in drinking or feeding in view of the diversity of both cholinergic and noradrenergic pathways (see below), and in view of the whole sequence of neurons which must be involved in controlling all the sensory, motivational and motor processes that are involved in drinking behaviour. The evidence on how pharmacological agents which affect neurotransmission in these systems in the brain influence drinking is considered below.

Another way in which pharmacological agents can stimulate drinking is by acting through peripheral mechanisms. As described below, it is likely, for example, that drinking elicited by the β-adrenergic agonist isoproterenol (isoprenaline) in the rat depends at least partly on the release of renal renin. Other substances, such as angiotensin, may stimulate drinking because of an action on special receptors which allow systemic angiotensin to influence brain processes.

Cholinergic drinking

The diversity of cholinergic pathways in the brain is indicated in figure 8.1 (Lewis & Shute, 1978). These pathways are presumed to be cholinergic on the basis of the presence of the catabolic enzyme acetyl-cholinesterase, although final confirmation is required as this enzyme marks some neurons which are not cholinergic. Some major projections to note are a diffuse projection which ascends from the midbrain reticular formation to the cortex (not shown in figure 8.1), a ventral tegmental pathway ascending from the substantia nigra and ventral

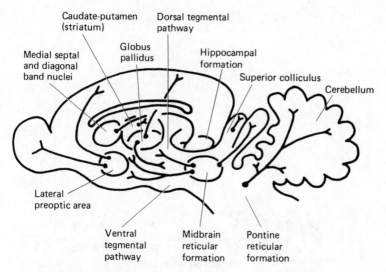

Fig. 8.1. Some cholinergic pathways in the brain. (Tentative identification based on content of acetylcholinesterase.) (From Lewis & Shute, 1978.)

tegmental area to the hypothalamus, thalamus and forebrain, and a projection from the basal forebrain nucleus of Meynert in the lateral preoptic area, lateral hypothalamus and substantia innominata to all areas of the cerebral cortex (Kievit & Kuypers, 1973; Divac, 1975; Lewis & Shute, 1978).

The first description of cholinergically induced drinking in the rat was by Grossman (1960, 1962a). He placed crystalline acetylcholine or carbachol in estimated doses of 1–5 μg into the lateral hypothalamus through a previously implanted guide cannula, and found that drinking started 4–8 min later and continued for 30–40 min, during which time about 12 ml of water was taken. Carbachol is a direct cholinergic agonist, and is often used in these studies in preference to acetylcholine because it is not metabolized by acetylcholinesterase and is thus not rapidly destroyed. The effect of the stimulation must have been central, i.e. on the brain, in that peripheral cholinergic stimulation depresses drinking. Other pharmacological agents used as control substances did not elicit drinking. An example of a dose–response curve for carbachol-induced drinking is shown in figure 8.2. As in most subsequent studies, the drug was in solution. It is clear that a dose as low as 2.7×10^{-10} moles (49 ng) elicited a reasonable drinking response, and that the response was behaviourally specific in that eating was not elicited (Miller, Gottesman & Emery, 1964). This cholinergic response was

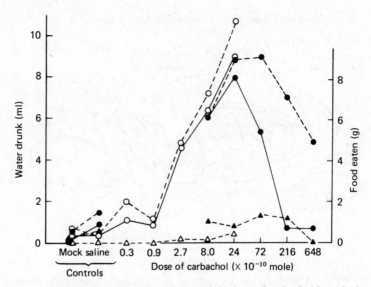

Fig. 8.2. Elicitation of drinking by injections of carbachol into the lateral hypothalamus of the rat (circles). Very little food intake (triangles) was elicited. The results from different experiments are indicated by solid and dashed lines. At the highest doses of carbachol, side effects prevented drinking. (After Miller, Gottesman & Emery, 1964.)

muscarinic in that a similar response was obtained with muscarine but not nicotine, and in that the response was blocked by atropine sulphate (Stein & Seifter, 1962). Injections of physostigmine, which prevents the destruction of endogenously produced acetylcholine, caused some drinking (Grossman, 1962b; Winson & Miller, 1970) or enhanced water-deprivation-induced drinking (Miller, 1965). This is consistent with the possibility that, when cholinergic substances are injected into the lateral hypothalamus and induce drinking, they act on an endogenous acetylcholine-producing system involved in drinking.

The brain sites in which carbachol injections can induce drinking in the rat are widespread, including, in addition to the hypothalamus, many parts of the limbic forebrain and limbic midbrain, such as the hippocampus, septal region, the mammillary body–interpeduncular region, the anterior thalamus, and the cingulate gyrus (Fisher & Coury, 1962, 1964). Diffusion or leakage of the carbachol to a common site of action is a possibility, and it is notable that these sites are distributed near the ventricles. With lower doses of carbachol (7.5 ng in 0.1 μl) the sites were less widespread and were adjacent to the third ventricle in the medial hypothalamus (Swanson & Sharpe, 1973). In fact, it appears that one site of action for carbachol-induced drinking is the subfornical organ, which is situated in the third ventricle, for the latency to the onset of drinking was shorter and the amount of water consumed was greater if the carbachol was applied to the SFO itself, rather than to the lateral hypothalamus or into the ventricle (Simpson & Routtenberg, 1972). Consistent with this was the finding that lesions of the SFO greatly reduced drinking in response to carbachol injected into the third ventricle (Simpson & Routtenberg, 1972). Although the SFO is thought to be one site with receptors which mediate angiotensin-induced drinking (see Chapter 4), these receptors appear to be separate from those for cholinergic drinking in that, in the SFO, carbachol-induced drinking was blocked by atropine and angiotensin-induced drinking was blocked by saralasin (p113), but not vice versa (Mangiapane & Simpson, 1979). Thus at this level, the cholinergic and angiotensin systems appear to be in parallel. Interestingly, although blockade of either system alone has little effect on water-deprivation-induced drinking, combined blockade (with atropine and saralasin) does attenuate water-deprivation-induced drinking (Hoffman *et al.*, 1978; see Chapter 4). Another possible site of action of carbachol is in the anteroventral third ventricle, perhaps in the organum vasculosum of the lamina terminalis which is, like the SFO, a circumventricular organ receiving a cholinergic input (Buggy & Fisher, 1976; Buggy, 1978).

Although there thus appears to be a cholinergic mechanism through which drinking can be elicited in the rat, one great peculiarity is that it is very difficult to obtain similar results in any other species. Central cholinergic stimulation in the cat can elicit rage, attack or sleep, but not drinking; in the monkey, it can block feeding and drinking; in the rabbit it can cause some eating; in the ring dove it does not elicit eating or drinking (see Fitzsimons, 1979); and in the dog, injections of 1 μg of carbachol into the third ventricle can cause some drinking (Ramsay & Reid, 1975).

Adrenaline and noradrenaline

Adrenaline (epinephrine) and noradrenaline (norepinephrine) can both influence drinking. Although adrenaline activates largely β-adrenergic receptors, and noradrenaline activates largely α-adrenergic receptors, this separation is not at all absolute. It is important in analysing adrenergic effects on drinking to separate clearly β- from α-adrenergic effects, for a β-adrenergic activation can lead to drinking via a peripheral site of action, and α-adrenergic activation of a system in the brain can inhibit drinking, as discussed below.

First, β-adrenergic stimulation, with for example isoproterenol (isoprenaline), can stimulate drinking (Lehr, Mallow & Krukowski, 1967). The drug was given peripherally, subcutaneously, and the effect was blocked by β-blockade with propranolol. Although Leibowitz (1971) found that drinking could be elicited by intrahypothalamic injections of isoproterenol, the doses were enormous (40 μg of the bitartrate) compared to the peripherally effective dose of several microgrammes (Lehr, Mallow & Krukowski, 1967), and probably acted by leakage to the periphery (Ramsay, 1978). Some evidence suggests that the β-adrenergic stimulation induces drinking by activating the renal renin–angiotensin system. β-Adrenergic stimulation with isoproterenol causes a dose-related increase in plasma levels of angiotensin II in the rat (Johnson *et al.*, 1981). Also, in both the rat and the dog bilateral nephrectomy (Houpt & Epstein, 1971; B. J. Rolls & Ramsay, 1975) and the competitive angiotensin inhibitor saralasin acetate (Chiaraviglio, 1979; B. J. Rolls & Ramsay, 1975) can abolish the dipsogenic effect of isoproterenol. An alternative to activation of the renin–angiotensin system is the possibility that, since isoproterenol causes a drop in blood pressure, it could stimulate drinking through vascular receptors. This idea is supported by the finding that, although nephrectomy abolishes drinking in response to low doses of isoproterenol, larger doses which cause greater reductions of blood pressure do stimulate drinking in rats (Hosutt, Rowland & Stricker, 1978) and dogs (Ramsay, 1978). Thus the renin–angiotensin system appears to be important for the initiation of drinking when blood pressure starts to fall, but as the reduction in pressure becomes more severe, activation of vascular receptors may also lead to drinking.

Secondly, α-adrenergic stimulation can inhibit drinking, by a direct action in the brain. The effect can be obtained with injections of 0.9 ng (5×10^{-12} moles) of noradrenaline into the rat hypothalamus and is blocked by α-adrenergic antagonists (Leibowitz, 1975a, b, 1981). At higher doses (e.g. 1–4 ng) noradrenaline injected into the region of the paraventricular nucleus of the hypothalamus of the rat elicited feeding with a typical latency of 4–5 min, and at higher doses (5–17 ng) this was usually preceded by a 1–3 min bout of drinking which started with a latency of less than 1 min after the injection. The injections also produced antidiuresis with a very short latency.

Dopamine

The nigrostriatal, meso-limbic, and meso-cortical dopamine pathways have been described in Chapter 7 (see figure 7.2). Intracranial injections

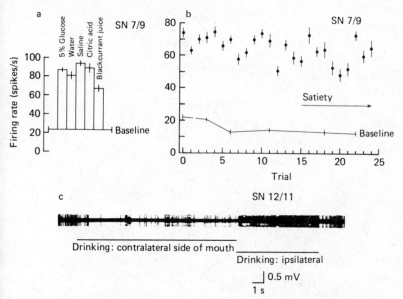

Fig. 8.3. (a) This neuron in the substantia nigra (SN 7/9) responded in relation to mouth movements made during feeding or drinking. Its firing rate increased independently of whether water, glucose, etc., was tasted, as long as the mouth movement was made. (c) A similar neuron in the substantia nigra (SN 12/11) with activity related to mouth movements responded when the monkey drank from a syringe when it was in the mouth ipsilateral but not contralateral to the recording site, and (b) responded by increasing the firing rate above the baseline line when the monkey made movements even after he had been fed to satiety. (From Mora, Mogenson & Rolls, 1977.)

of dopamine in the range of 32.5–260 nmol are relatively ineffective in eliciting drinking, although blockade of dopamine receptors with haloperidol or spiroperidol reduced drinking induced by angiotensin or water deprivation but not drinking induced by carbachol or cellular dehydration (Setler, 1973, 1977; Fitzsimons & Setler, 1975; Fitzsimons, 1979; Leibowitz, 1981). Thus there is some evidence that a dopaminergic system is necessary for some stimuli of drinking to function, but the pathway involved and its function are not at all clear.

As described in Chapter 7, destruction of the dopaminergic nigrostriatal bundle results in adipsia (lack of drinking) and aphagia (lack of eating) and a sensory-motor disturbance in which there is failure to orient to environmental stimuli (e.g. Marshall, Richardson & Teitelbaum, 1974). The adipsia could arise because a specific control of thirst and drinking has been disrupted, or it could arise as a side effect because of destruction of a system normally involved in, for example,

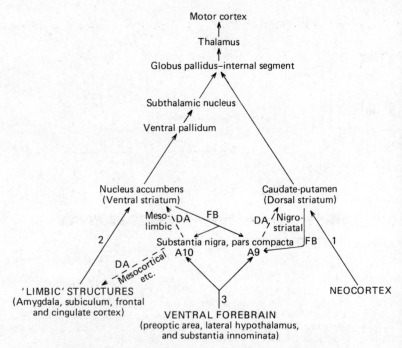

Fig. 8.4. The anatomical situation of dopamine (DA) pathways. The nigrostriatal dopaminergic system is in a position to influence transmission through the caudate and putamen, which receives inputs from the neocortex (1) and projects to motor structures, primarily via the globus pallidus. The mesolimbic and mesocortical dopaminergic systems are in a position to influence transmission through the ventral striatum (nucleus accumbens), which receives inputs from 'limbic' areas (2) such as the frontal and cingulate cortex and the amygdala, and projects to motor structures as shown. Dopaminergic activity itself is under feedback control as shown (FB), and is influenced by other inputs, from, for example, ventral forebrain systems (3) including the lateral hypothalamus and preoptic area.

movements and their initiation, whether for drinking or not. To investigate the function of the nigrostriatal system in drinking and feeding, recordings have been made from single neurons in the substantia nigra during drinking and feeding in the monkey (Mora Mogenson & Rolls, 1977). Some nigral neurons responded during feeding and drinking, but the responses were associated with movements of, for example, the mouth or arm, occurring equally during feeding and drinking and regardless of whether or not the monkey was hungry or satiated, provided that the same movement was made (Mora, Mogenson & Rolls, 1977) (figure 8.3). Thus, the function of this system appears to be related to the initiation and control of movements, rather than more specifically to the control of thirst.

The function of the nigrostriatal system can be assessed more fully by considering the connections and function of the striatum as a whole. The caudate nucleus and putamen, the parts of the striatum that are influenced by the nigrostriatal dopaminergic pathway, receive their main projection from all areas of the cerebral cortex and, in turn, influence the motor cortex (via the globus pallidus and ventral thalamic nuclei) (see figure 8.4). The dopaminergic nigrostriatal bundle is thus anatomically in a position to influence transmission from all areas of the cortex through the caudate nucleus and putamen to the motor cortex (figure 8.4). The neuronal responses in the caudate nucleus and putamen in general reflect the responses found in the part of the cortex from which each given part of the caudate or putamen receives its projections. Thus, for example, there are visual responses, which habituate, in the tail of the caudate nucleus, there are responses in the head of the caudate to environmental stimuli or to cues which the animal uses in the preparation for and initiation of behavioural responses such as eating or drinking, and in the putamen there are neurons with responses related to movements of, for example, the arm or mouth (E. T. Rolls, Thorpe, Maddison *et al.*, 1979; E. T. Rolls, Perrett *et al.*, 1979; E. T. Rolls, 1980, 1981a, b). Given that dopamine influences the responses of many of the neurons as shown by micro-iontophoretic application of dopamine to single neurons in the behaving monkey (E. T. Rolls, Thorpe, Perrett, Boytim, Ryan & Szabo, in preparation), it is consistent that animals with damage to the nigrostriatal system fail to orient to environmental stimuli, and fail to initiate movements and behaviour such as eating and drinking (E. T. Rolls, 1981a). The neurophysiological analysis thus suggests that the nigrostriatal dopaminergic system is important in influencing transmission of information from different areas of the cerebral cortex to motor structures, and that this information is used in the preparation of and initiation of behavioural responses. Adipsia may thus follow damage to this system because of difficulty in initiating behavioural responses, rather than because this system contains any specific control system for thirst and drinking. However, control signals for drinking could enter the striatum in a number of ways (e.g. via the cortex, or via the hypothalamic projections to the nigrostriatal and meso-limbic and meso-cortical dopamine systems), and the striatum does appear to be part of a system necessary for the final expression of behaviour, including drinking behaviour.

Summary

Cholinergic stimulation of the brain induces drinking, but only in the rat. This cholinergic effect on drinking can be separated pharmacologically using receptor blocking agents from the system described in Chapter 4 for angiotensin-induced drinking. Peripheral β-adrenergic stimulation elicits drinking which depends on the presence of the renin–angiotensin system. Central α-adrenergic stimulation can inhibit drinking. The dopaminergic nigrostriatal system is involved in the preparation for and initiation of behavioural responses, including drinking.

9   Problems of thirst and fluid balance

The fluid intakes and the fluid requirements of healthy individuals can vary enormously. Some tend to be moderate drinkers while others always drink copiously. Even within an individual intake can show wide fluctuations depending on the climate, social situation, availability of fluids, etc. Normally the variability in fluid intake is of little consequence because the kidneys can efficiently eliminate excess water in a dilute urine or can concentrate the urine to conserve water. Sometimes, however, fluid intake is so excessive or so low that it becomes a clinical problem. A good example of the dramatic nature of aberrations of thirst comes from the account in 1856 by Atkinson of 'A remarkable case of intemperate drinking'.

He was active and industrious, and enjoyed good health. He complained of nothing but excessive thirst. To such a degree did he suffer from this cause, that it was hard to resist the conviction that he had been bitten by the 'Dipsas', a serpent known among the ancient Greeks, whose sting produced a mortal thirst.

Although a sober man, he was the most intemperate drinker I ever knew, from four to six gallons of water being required to keep him comfortable during the night, while his daily ration of this, to him literal 'aqua vitae', amounted to not less than from eight to twelve gallons. He always placed a large tubful near his bed, on retiring for the night, which often proving insufficient, he was forced to hurry to the spring to obtain relief from the intense suffering occasioned by the scanty supply. He has frequently driven the hogs from mud-holes in the road, and slaked his thirst with the semi-liquid element in which they had been rolling, himself luxuriating in that which had afforded only a moderate degree of enjoyment to the swine.

The factors which stimulate drinking in man have been mentioned throughout the book where appropriate. It seems clear that depletions of either the cellular or the extracellular fluid compartment can lead to fluid intake. Thus, in conditions where body water or electrolytes are lost thirst will appear as a symptom. Some examples of such conditions are severe vomiting or diarrhoea (occurring for example in cholera), impairment of the release of antidiuretic hormone, and some forms of kidney disease which lead to excessive loss of water or electrolytes. Sometimes thirst can occur when there are no fluid depletions, for example in compulsive water drinking attributed to psychological disturbance. Also, in some forms of brain disease there may be excessive drinking because of irritation of areas of the brain concerned with drinking, and in some kidney diseases elevated renin and angiotensin levels lead to polydipsia. In these examples the drinking is inappropriate for the maintenance of fluid balance and, if the kidneys cannot cope with the excessive load, it can be considered pathological in that dangerously low plasma electrolyte concentrations may be found, or if both salt and water are retained oedema may develop. Several examples of polydipsia will be discussed in more detail and then decreased thirst and severe fluid deprivation will be considered.

Examples of polydipsia

Diabetes mellitus and diabetes insipidus

Greek and Roman physicians used the term diabetes to refer to diseases characterized by the production of a copious urine. Diabetes mellitus, in which there is insufficient insulin for the cellular uptake of glucose, is accompanied by polyuria because some of the excess circulating glucose is lost in the urine. By osmosis the glucose draws water into the urine (the normal reabsorption of water in the kidney is hindered) and large amounts of water are lost in the osmotic diuresis that ensues. The resulting dehydration leads to intense thirst.

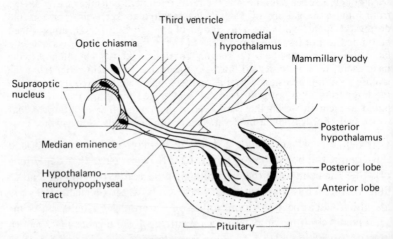

Fig. 9.1. The supraoptic–posterior pituitary system. Antidiuretic hormone is formed in the supraoptic nucleus and travels along the hypothalamoneurohypophyseal tract to be released by the posterior lobe of the pituitary gland into the bloodstream. Damage to this system leads to diabetes insipidus which is characterized by polyuria (excessive urination) and polydipsia (excessive drinking).

Diabetes insipidus is a disorder in which the main symptoms are excessive urine output and excessive thirst. The disease is often due to damage anywhere along the supraoptic–posterior pituitary system (see figure 9.1) which is involved in the release of antidiuretic hormone. This lack of ADH leaves the kidneys unable to concentrate the urine (Chapter 2). Because both the polydipsia and the polyuria are controlled by administration of ADH, it has been supposed that the polydipsia is simply secondary to the excessive loss of urinary water. Support for this view came from an experiment (Richter & Eckert, 1935) in which the polydipsia which resulted from removal of the posterior lobe of the pituitary was eliminated by tying both ureters to prevent urine formation. Presumably when the kidneys are intact, loss of water in the urine leads to dehydration of the cellular and extracellular fluid compartments and drinking is a response to these depletions.

Body fluid analysis of rats with hereditary diabetes insipidus indicates that there is cellular dehydration (i.e. increased osmolality and sodium) and a reduction of plasma volume (i.e. increased haematocrit) and an increase in plasma angiotensin II levels (Haack *et al.*, 1975). Treatment of these rats with ADH restored plasma volume, and reduced angiotensin levels, and the polydipsia disappeared (although significant elevations of plasma sodium and osmolality remained). This indicates that hypovolaemia and increased angiotensin may contribute to the thirst in hereditary diabetes insipidus. This is supported by the finding that blocking the formation of angiotensin II reduced the drinking rates of rats with hereditary diabetes insipidus (Henderson *et al.*, 1979) but this may have been due to a reduction in renal water loss (Stamoutsos *et al.*, 1981). It cannot be assumed that the fluid changes in hereditary diabetes insipidus reflect those which occur in experimentally produced diabetes insipidus, since in the hereditary disorder there is atrophy of the anterior pituitary and the adrenal cortex which would lead to complex hormonal changes.

In the case of rats in which the diabetes insipidus results from damage to the hypothalamus it has been suggested that there is a disorder of thirst mechanisms rather than just of ADH release. Smith & McCann (1962) found that damage to the median eminence of the tuber cinereum which is in the hypothalamus gave rise to diabetes insipidus in rats with intact kidneys and also slightly increased the water intake in rats with both kidneys removed. Such excess drinking in rats with diabetes insipidus without kidneys (primary polydipsia) would be important if it could be shown to result from damage to a hypothalamic 'thirst satiety centre' responsible for stopping drinking when sufficient water had been drunk to satisfy the body's needs. B. J. Rolls (1970) has shown, however, that the primary polydipsia following median eminence lesions is due to transient irritation of the lateral hypothalamus caused by deposition of metallic ions from the lesioning electrode and not destruction of a thirst 'satiety centre'. With this transient irritation eliminated, nephrectomized rats with diabetes insipidus responded normally to both cellular and extracellular thirst stimuli.

Recently the argument for primary polydipsia in diabetes insipidus has been revived (Hennessy, Grossman & Kanner, 1977). Knife cuts in the posterior hypothalamus, which transect fibres entering and leaving the ventromedial hypothalamus, impair ADH release. It was suggested that not all of the excessive fluid intake observed in these animals was due to renal fluid loss. This could be the case but the critical experiment to prove primary polydipsia, namely to look at the drinking response of the animals after bilateral nephrectomy, was not performed. Thus, the question of whether primary polydipsia can follow lesions in the hypothalamus remains an open issue as does the broader question of whether there is a thirst 'satiety centre' (see Chapter 7).

There have been several clinical reports which have indicated that thirst is the primary disorder in diabetes insipidus. There was the case of the man kicked by a horse (see Chapter 3), described by Nothnagel (1881), who drank 3 l of water and beer within half an hour of his fall, but did not pass any urine for 3 h. Also, in a case of post-traumatic diabetes insipidus the intense thirst of a patient stopped abruptly after puncturing a subarachnoid cyst at the base of the brain which had raised the hypothalamus and stretched the hypophysial stem (Kourilsky, 1950). More recently Stuart, Neelon, and Lebovitz (1980) studied water metabolism and thirst in patients with hypothalamic–pituitary sarcoidosis (a disease characterized by multiple microscopic nodules or granulomas especially in the subependymal hypothalamus around the third ventricle). Of the seven patients found to have abnormal water metabolism three had excessive thirst but normal ADH function (i.e. primary polydipsia), while one patient was found in each of the following four categories: deficient thirst and normal ADH function, deficient thirst and severe ADH deficiency (i.e. diabetes insipidus), normal thirst and partial ADH deficiency, and normal thirst and severe ADH deficiency. Thus disorders of thirst and ADH release can occur both separately and in combination. Anatomical and histological details of the brains of the patients mentioned above are not available, but it is possible in cases suggestive of primary polydipsia that the excessive intake could be due to continued irritation of brain areas concerned with initiating thirst rather than to damage to areas concerned with inhibiting drinking.

Compulsive water drinking

Sometimes excessive fluid intake, which tends to vary widely from day to day, is seen in patients with normal or even elevated levels of ADH. Compulsive water drinking (psychogenic polydipsia or potomania) may therefore be inappropriate to fluid needs and may lead to dangerously low plasma sodium levels and hyponatraemic convulsions. The cause of the polydipsia is not clear, but the syndrome is usually found in patients with psychiatric disorders, particularly middle-aged women. Some of them may have developed an obsession for water while in others there could be a physical explanation for the thirst. There are several reports that show that compulsive water drinking can be associated with hydrocephalus (Hogan & Woolsey, 1967; Peterson & Marshall, 1975). This could then be another example of drinking resulting from chronic irritation of areas of the brain concerned with the initiation of drinking since an enlargement of the third ventricle is seen. Such irritation could also lead to excessive release of ADH which may exacerbate the condition. The inappropriate release of ADH may also be caused by emotional factors such as stress or pain. For example, an apparently normal child with severe toothache and polydipsia developed hyponatraemic convulsions, presumably because of pain-induced release of ADH (Pickering & Hogan, 1971). Acute psychotic attacks with polydipsia are sometimes correlated with elevated ADH (Dubovsky, Grabon, Berl & Schrier, 1973). In some psychiatric disorders drug therapy may affect renal function and this can lead to a dangerous condition in compulsive drinkers. For example, we (J. Ledingham, B. J. Rolls & J. Gibbs) observed a patient who had reduced renal output as a result of phenothiazine treatment for schizophrenia. This patient compulsively drank tea – up to 100 cups a day. During these compulsive episodes the patient developed hyponatraemic convulsions. The behaviour was apparently obsessional rather than due to true thirst in that the patient would only reluctantly drink the water required for free water clearance studies but avidly drank tea. This emphasizes the importance of palatability in the controls of fluid intake (see Chapter 5).

Compulsive water drinking can sometimes be difficult to diagnose. The thirst is more variable than that seen in diabetes insipidus, and the patient's plasma is usually hypotonic (i.e. plasma osmolality below 270 mosmoles/kg H_2O), whereas the patient with diabetes insipidus usually has hypertonic plasma (plasma osmolality above 295 mosmoles/kg H_2O). Also, the compulsive water drinker should respond to fluid deprivation by releasing ADH and concentrating the urine (although chronic overdrinking may impair this ADH response for a few days), whereas the diabetic patient will not. These distinctions do not always hold. Further diagnostic tests can be found in Schrier & Berl (1976).

Beer potomania

If excessive quantities of beer are consumed (beer potomania) and most of a person's energy is derived from the beer, dangerous dilution of the blood is sometimes found. In normal persons the electrolytes and protein derived from food help to maintain the volume and composition of the plasma and up to 15 l of water can be consumed without the development of hyponatraemia. If beer is the only source of food and fluid, as may be the case in beer potomania, hyponatraemia may develop with an intake of only 6 l/day. This is because the loss of electrolytes and urea in the urine are likely to exceed the amounts consumed. A remedy then for beer potomania is that the beer should be salted (Fanestil, 1977)!

Symptoms of hyposmolality or hyponatraemia

Several examples of situations in which plasma dilution may occur are discussed above. Such dilution may be detected clinically by measuring plasma sodium concentration or osmolality, but the question naturally arises as to whether such dilution can be detected readily by observing a patient's symptoms. The symptoms of hyponatraemia are non-specific and variable. A patient may have no symptoms, mild symptoms (confusion or headaches), moderate symptoms (nausea and vomiting), or severe symptoms (convulsions, coma, death). These symptoms tend not to be correlated with the degree of hyponatraemia or hyposmolality, but tend to be associated with rapid decreases in plasma concentration, The symptoms are usually attributed to cerebral oedema (Fanestil, 1977).

Lithium treatment for manic-depressive psychosis

Various pharmacological agents have been found to impair the renal capacity to produce a concentrated urine (Schrier & Berl, 1976). Here we will discuss the effects of lithium, which may affect both fluid intake and output. Lithium carbonate is a frequently used prophylactic treatment for manic-depressive psychosis. Polyuria and polydipsia may accompany such treatment. Lithium can affect urine-concentrating mechanisms and may lead to nephrogenic diabetes insipidus (i.e. due to renal impairment) which is little improved by ADH administration. Of particular interest is the evidence which indicates that lithium may directly affect thirst mechanisms and cause a primary polydipsia. Since patients treated with lithium may complain of thirst because of their psychiatric disorder (possibly manifested as compulsive water drinking), it is difficult to know if the thirst is primary or secondary to increased urinary water loss. However, several animal studies indicate that lithium may directly affect thirst. In rats the drinking may precede increases in urine flow and occurs even when the ureters have been ligated. Increased thirst following lithium can occur in rats without ADH (i.e. with hereditary diabetes insipidus), and in normal rats when there has been no change in osmolality or plasma volume (see Forrest & Singer, 1977; Jenner & Eastwood, 1978). Taken together, these results indicate that lithium can directly affect thirst mechanisms. The cause of such primary thirst is not clear. The thirst was blocked by haloperidol and propranolol but not by atropine, and therefore appears to depend on dopaminergic and β-adrenergic pathways. It has been suggested that the renin–angiotensin system may be involved in the primary polydipsia. Elevated plasma renin levels may be found when high doses of lithium are administered, but since this elevation in renin could be due to hypovolaemia resulting from urinary sodium and water loss, it does not necessarily provide an explanation for the primary polydipsia. It has been suggested that a unique homeostatic mechanism may be involved in the increased drinking accompanying lithium treatment. Lithium can cause feelings of sickness, and in rats lithium chloride is frequently used to produce learned aversion. The increased water intake of rats given lithium helps to speed the renal elimination of lithium. Animals may learn quickly that increased fluid intake hastens the detoxification of the body. This phenomenon has been called 'antidotal thirst' (D. F. Smith, Balagura & Lubran, 1970).

Although many patients receiving lithium will experience transient thirst and polyuria, gross disturbances of fluid balance and renal function are prominent complications in only about 10% of those treated. When these complications do occur, however, they can be dramatic. For example, we (J. Ledingham, B. J. Rolls & R. Wood) recently saw a patient who had been receiving lithium therapy for 7 years. Her manic-depressive condition was well controlled, but she had always experienced some thirst with the drug. After about 6.5 years she experienced a dramatic increase in thirst which she claimed preceded any change in urine flow. Her life now centres around her thirst. She claims to drink 30 pints a day of anything she can get except alcohol. The fluid must be gulped rapidly. Swilling fluid in the mouth gives no relief. Because much of the fluid is high in calories, she no longer eats. She has trouble sleeping because she has to urinate about six times in the night. Such patients may be found to have diabetes insipidus due to damage to the nephrons. This disorder usually reverses upon withdrawal of lithium therapy, and it is not clear there there is any permanent damage to the kidneys (Jenner & Eastwood, 1978).

Renal hypertension

Thirst has been noted clinically in various conditions associated with elevated plasma renin, such as normal pregnancy, anorexia nervosa and malnutrition, haemorrhage, and sodium deprivation. One of the most dramatic examples of thirst associated with elevated renin occurs in renal hypertension (hypertension generally refers to an elevation of both the systolic and diastolic components of the blood pressure). J. J. Brown *et al.* (1969) described a patient with renal hypertension, in whom urine output was impaired and gross oedema developed, indicating inappropriate fluid intake. Plasma sodium was low but plasma renin was elevated and the patient developed excruciating thirst leading to illicit attempts to obtain water. He confessed that he had concealed and later drunk water which had been used for bathing and shaving. The hypertension was so severe and uncontrollable that this patient had to be nephrectomized bilaterally. Removal of the kidneys reduced renin 200-fold and abolished the thirst before changes had occurred in sodium and water balance. In another hypertensive patient with elevated plasma renin, bilateral nephrectomy was performed to relieve the constant thirst (Rogers & Kurtzman, 1973).

Renal hypertension can be produced experimentally in various ways and in various animals. In the dog reducing the diameter of one renal artery and removing the opposite kidney lead to hypertension (the one-kidney Goldblatt model) and a dramatic increase in water intake and a change in the pattern of drinking over the first few days after constriction (see figure 9.2). In this situation the drinking is likely to be due to elevated angiotensin, since the salt and water retention which are occurring make it unlikely that hypovolaemia or cellular dehydration could be the thirst stimulus.

In 1969 Fitzsimons showed that constricting the renal arteries of rats increased water intake. The elevation of intake was correlated with the rise of plasma renin (Leenen, Stricker, McDonald & De Jong, 1975) which also promoted fluid retention by stimulating ADH release. The elevation of water intake occurred despite lowered plasma sodium and may have contributed to the elevation of blood pressure seen during the development of hypertension. This view is supported by the finding that restricting fluid intake reduced the degree of hypertension that was reached in rats with both renal arteries constricted (Leenen, de Jong & de Wied, 1972).

The studies of Leenen and colleagues concerned rats in which

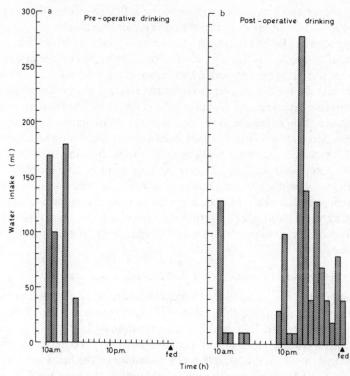

Fig. 9.2. Typical patterns of drinking by one dog (a) before and (b) after the experimental production of renal hypertension. (From B. J. Rolls & Ramsay, 1975.)

hypertension was produced by constriction of both renal arteries. Constriction of only one renal artery, leaving the other kidney intact (two-kidney Goldblatt model), leads to the development of hypertension which initially is similar to that just described and which is characterized by polydipsia, high renin, and sodium retention. In the initial, benign, phase plasma volume increases. As the syndrome develops, if a critical level of blood pressure is exceeded, salt and water are lost and the renin–angiotensin system is further activated. This marks the onset of the malignant phase of hypertension. The animals remain thirsty and this could be due to high renin, hyperosmolality, and hypovolaemia. In this phase the condition of the rats deteriorates rapidly. If these animals are offered isotonic saline to drink, they show compulsive saline drinking and a dramatic improvement in their condition. Not only is saline consumption therapeutic in malignant hypertension, it can also be preventive. Rats which were developing malignant hypertension lowered the blood pressure by voluntarily selecting saline to drink (Möhring *et al.*, 1975). This salt appetite could be due either to elevated aldosterone or angiotensin. Before deciding whether salt would be beneficial, it is essential to assess the volume of the extracellular fluid compartment, since in non-malignant hypertension retention of salt and water would exacerbate the condition as hypervolaemia elevates the blood pressure.

Thus, there is a marked elevation of fluid intake in renal hypertension and this is likely to be due to the higher levels of circulating angiotensin (Mann, Johnson & Ganten, 1980) since volume and pressure receptor input is elevated and this would normally inhibit drinking. Angiotensin and the increased fluid intake may be important in the aetiology of hypertension in that lesions in the region of the anteroventral third ventricle, which is thought to be a site for angiotensin-induced drinking (see Chapter 4) and pressor responses, prevented the development of experimental renal hypertension in rats (Buggy, Fink, Johnson & Brody, 1977).

Congestive cardiac failure and thoracic caval constriction

Holmes (1960) has reported that patients with congestive cardiac failure sometimes show intense thirst. There is often considerable oedema in these patients so the fluid intake is inappropriate to fluid needs. In man congestive cardiac failure is a varied syndrome. The term 'cardiac failure' or 'heart failure' simply refers to failure of the heart muscle so that the heart is unable to pump sufficient blood to the peripheral tissues (Davis, 1967). When heart failure is accompanied by an abnormal increase in blood volume and interstitial fluid volume (oedema), and venous congestion ensues, the condition is known as congestive cardiac failure.

The reasons for the increased thirst and inappropriate fluid intake which lead to oedema in congestive heart failure have not been established, although arterial constriction and elevated plasma angiotensin levels may be involved. A representative and stable animal model of heart failure is difficult to produce, but much of the exploratory work on the mechanisms of heart failure has been done in dogs in which the inferior vena cava is partially constricted close to its entry to the heart. Although this situation is not exactly comparable to congestive cardiac failure in that there is no defect of the heart muscle and the venous congestion occurs below the constriction and not immediately around the heart, many of the mechanisms of salt and water metabolism in these dogs appear to be similar to those in experimental heart failure (Davis, 1967).

In the dog such constriction led to elevated and inappropriate fluid

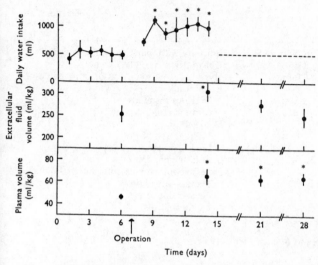

Fig. 9.3. The plasma volume, extracellular fluid volume and daily water intake before and after thoracic inferior vena caval constriction. On days 8–14 the dogs ($N = 4$) were given *ad libitum* access to water. The first two weeks of fluid-intake restriction are shown. On days 15–28, the dogs were restricted to their mean pre-operative water intake. This restriction reduced extracellular fluid volume to pre-operative levels and greatly improved the condition of the dogs. Asterisks indicate results that are significantly different from the pre-operative level. (From Ramsay, Rolls & Wood, 1975.)

intake with marked oedema. This elevated intake was likely to be due in part to increased plasma angiotensin since administration of the competitive angiotensin inhibitor, saralasin acetate, reduced fluid intake. If the elevated intake persisted the condition of the dogs deteriorated, but, if the fluid intake was at this stage restricted to pre-operative levels (see figure 9.3), the condition of the dogs markedly improved and the oedema diminished (Ramsay, Rolls & Wood, 1975). Thus, as was found with renal hypertension, fluid intake can be important in the aetiology and development of symptoms associated with the disease state. These findings could be of clinical importance and imply that there should be clinical assessment of the role of water and salt intake in the development of some disease states.

Dehydration

Hyperosmolality may be encountered in a variety of clinical situations, and is usually associated with increased plasma concentrations of

Table 9.1. *Causes of hypernatraemia*

Children	Adults
Gastroenteritis with diarrhoea	Infirmity with inability to obtain water
Salt intoxication	Dehydration secondary to hot climate
Improper mixing of formula	Hyperalimentation – oral or intravenous
	Renal water loss
	Diabetes insipidus
	Chronic renal failure
	Dialysis
	Malfunction of proportioning system
	Loss of water
	Steroid hypersecretion
	Excessive intravenous infusion of sodium
	Therapeutic abortion
	Postcardiac arrest
	Lactic acidosis

From Arieff, Guisado & Lazarowitz (1977).

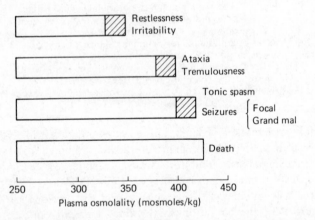

Fig. 9.4. Effects of acute hyperosmolality on symptomatology in experimental animals. Cross-hatched areas represent the range of plasma osmolality for each group of manifestations. (From Arieff, Guisado & Lazarowitz, 1977.)

sodium, urea, glucose, or ethanol. Clinically hypernatraemia is most frequently seen in the very young or very old. The most common causes of elevated plasma sodium are shown in table 9.1. Hyperglycaemia is most often seen in patients with uncontrolled diabetes mellitus. Elevated plasma urea is most often associated with renal failure. The effects of acute hyperosmolality in animals are summarized in figure 9.4.

Hypernatraemia

A common cause of hypernatraemia is an impairment of thirst mechanisms. In experimental animals some hypothalamic lesions, for example in the lateral hypothalamus (see Chapter 7), impair thirst and hypernatraemia ensues if the animals are not maintained with intubated water. In man, hypernatraemia is associated with a variety of brain lesions and often occurs because an impairment of sensory and motor functions makes it difficult to obtain adequate amounts of water. 'Essential hypernatraemia' has also been reported in patients with normal thirst mechanisms. It appears that central osmoreceptors have been reset or changed in sensitivity in that plasma sodium is maintained at a high level even when patients are forced to drink extra water. In most of these patients ADH and renal mechanisms are apparently functioning normally.

Hypernatraemia can be very serious in humans and is associated with high mortality. In children with either acute or chronic hypernatraemia neurologic abnormalities and abnormal electroencephalograms are common. In adults the symptoms are quite variable, but neurologic abnormalities and seizures are found. It is, however, difficult to determine whether such symptoms are due to the hypernatraemia or the precipitating disorder. Because hypernatraemia is often associated with seizures, the changes in brain water and electrolyte content have been studied in experimental animals. When the hypernatraemia lasted for only several hours there was an increase in brain osmolality and sodium concentration and a decrease in brain water and cellular dehydration. Normally such changes would be potent thirst stimuli. An interesting change took place if the hypernatraemia lasted for more sustained periods (i.e. more than several hours and up to 7 days). In this situation brain water content remained at normal levels (and intracellular volume was maintained intact) (Arieff, Guisado & Lazarowitz, 1977). As cellular dehydration is a critical thirst stimulus this could imply that, during severe hydration, thirst would peak early in the deprivation.

Many of the most graphic accounts of dehydration are given in Wolf (1958). He reviews accounts of survivors of long periods of dehydration due to being stranded in the desert or at sea. These accounts make very poignant the suffering which accompanies dehydration, but since few measurements were made in such situations it is difficult to draw firm conclusions about critical levels of depletion. During World War II the fighting in the deserts of North Africa made it essential that survival in the desert be studied. Adolph summarizes much of the evidence gathered

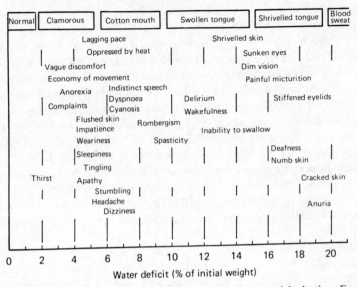

Fig. 9.5. Signs and symptoms characteristic of desert dehydration. Each sign is located at the water deficit at which it is usually first noted. (From Adolph, 1947.)

by scientific expeditions at that time in the classic book *Physiology of Man in the Desert* (1947). Interestingly, he notes that thirst was noticeable early in dehydration, but did not increase much in intensity as the water deficit increased. The order of appearance of the signs and symptoms of dehydration is characteristic (see figure 9.5). In the desert an average man can tolerate deficits of 3–4 % of body weight with only moderate impairments of efficiency; at 5–8 % he is fatigued, spiritless, complaining and prone to collapse; and at more than 10 % he is unco-operative and shows gross physical and mental deterioration. Accurate observations on man are not available with deficits greater than 11 %. The lethal point occurs somewhere between a 15 % and 25 % deficit, depending on the air temperature, exposure and speed of dehydration.

Conclusions

Fluid intake can be quite variable and normally the kidneys can cope with fluctuations by modifying the volume and concentration of the urine. Sometimes fluid intake can be so high or so low that it becomes a clinical problem. Throughout the history of studies of thirst clinical observations have had an important influence. For example, studies of patients with diabetes insipidus (copious drinking and urination) led to the suggestion that the excess drinking might be due to destruction of an area of the brain responsible for terminating drinking. Although the existence of such a 'satiety centre' has not been demonstrated experimentally, the studies have been useful in focussing attention on the role of the central nervous system in thirst. Thirst has been noted clinically in various conditions associated with elevated plasma renin, for example in renal hypertension and congestive heart failure. Experiments on animal models of these conditions have supported the idea that angiotensin can stimulate fluid intake, and that fluid intake can be important in the aetiology and development of some disease states. Not all drinking which is seen clinically can be related to an identified physiological imbalance, such as fluid depletion, elevated angiotensin, or brain stimulation. This leads on to the broader question of whether drinking can always be defined in terms of physiological need states. This question is considered in Chapter 10.

In previous chapters, our main concern has been to consider drinking as a response to changes in the body fluids. In this chapter we consider whether 'normal' drinking is in response to body fluid changes. We discuss some types of drinking which are clearly not in response to changes in the body fluids. Then we review the extent to which body fluid changes do account for drinking in response to water deprivation. Finally we consider the question whether spontaneous drinking, with water and food freely available, is in response to body fluid changes.

First, to introduce this problem of whether drinking is in response to the body fluid changes which can be measured and manipulated in the laboratory, we can note that Bolles (1979) has put forward the sceptical view that studies employing traditional physiological techniques in the laboratory are little more than playing with a 'toy rat'. This toy rat has a device at the front end that switches on a fluid intake mechanism and another device at the back end that lets fluid out. If the eliminator is eliminated, the toy rat is little more than a semipermeable membrane separating two fluid compartments. Bolles implies that this toy rat has taught us little apart from the way in which membranes work and the way in which rats will drink if they are nothing more than watery spaces separated by membranes. This toy rat is thought too simple to be relevant to real drinking behaviour.

> The real rat, it seems, has a great variety of reasons for drinking. It drinks in connection with eating, and it drinks in anticipation of thirst. Sometimes it learns where water is in its environment, but sometimes it has to search for water. Sometimes, because the rat is as lazy as the rest of us, it will forego drinking if too much searching or working for water is required. It paces itself with a diurnal cycle, but it also goes on binges, drinking needlessly. The rat in the laboratory cage will drink just because water is there. It may drink just because it has finished a meal or even because it has finished a bite of food (Falk's phenomenon).
> The real rat is a marvellously complicated drinker. (Bolles, 1979.)

The question examined in this chapter is to what extent drinking occurs normally in response to specific body fluid deficits, and to what extent it is 'non-regulatory', 'non-homeostatic', or 'secondary', that is, drinking which apparently does not serve to repair a deficit but is in excess of need (Rowland, 1977; Toates, 1979). A brief consideration of what is meant by homeostasis is worthwhile. For the cells of the body to function

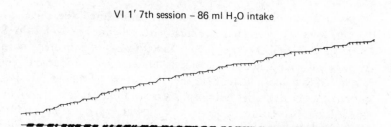

VI 1' 7th session – 86 ml H$_2$O intake

Fig. 10.1. An example of schedule-induced polydipsia. The rat pressed a bar for food pellets on a 1-min variable-interval schedule (VI 1'). Each bar press for food moved the upper pen upwards and the delivery of a food pellet produced a downward movement of the same pen. The lower line records every twelfth lick on a water spout. Characteristically, after a food pellet was earned there was a burst of licking. (From Falk, 1961.)

normally the extracellular fluid must remain relatively constant. There is, however, a continual exchange of water between the extracellular compartment and the environment. W. B. Cannon (1947) used the term 'homeostasis' to describe 'the various physiological arrangements which serve to restore the normal state once it has been disturbed'. Drinking which is not in response to body fluid deficits, and which in this sense is non-homeostatic (this sense is used commonly; e.g. Rowland, 1977; Bolles, 1979; Toates, 1979), is considered next.

Examples of drinking which is not in response to body fluid changes

Schedule-induced polydipsia (SIP)

Animals (rats, mice, pigeons, squirrels, rhesus monkeys and chimpanzees) which have been reduced to about 80% of their normal body weight and then given small pieces of food at intervals develop polydipsia known as schedule-induced polydipsia (SIP), psychogenic polydipsia, or adjunctive drinking behaviour (Falk, 1961, 1977). The drinking occurs in the interval between the delivery of the food pellets (figure 10.1), and in volume is about 10 times more than would be consumed if the food were given all at once. Rats will work to obtain water, indicating that the drinking is motivated. For example, food-deprived rats pressed a lever as many as 50 times to gain access to water after delivery of a food pellet (Falk, 1966), and they emitted more licks if the hole in the drinking tube was reduced, thus keeping the volume consumed constant (Freed & Mendelson, 1977).

Schedule-induced polydipsia is a good example of 'non-regulatory' drinking. It occurs when there is no specific fluid deficit, and is not greatly reduced by 10-ml preloads of water or saline (Porter, Young &

Moeschl, 1978). If normal renal function is impaired, the animals will become overhydrated, but eventually this hydration will stop SIP (Stricker & Adair, 1966). If renal function is not impaired, normal hydration is maintained and a copious dilute urine is produced (Kenny, Wright & Reynolds, 1976). SIP, then, does not serve homeostasis but on the other hand it does not normally upset fluid balance. It should be stressed that SIP occurs under very particular experimental conditions and is unlikely to be a condition which animals have ever encountered in their natural environment. Animals are therefore unlikely to have evolved mechanisms for coping with the situation, and the situation does not necessarily relate to the controls of normal drinking.

Agreement has not been reached on the cause of SIP but various theories have been put forward. Falk (1977) has suggested that SIP is equivalent to displacement activity which occurs in more natural situations and that it helps to stabilize other behaviours such as mating, intermittent feeding, etc. Staddon (1977) suggests that SIP is an adaptive behaviour which removes the animal from a situation of non-reinforcement and activates it to seek reinforcement elsewhere, for example from drinking. Both food deprivation and intermittent food induce arousal and it may be this arousal which leads to SIP. In this context it has been observed that two of seven schizophrenic patients sometimes drank 400–600 ml water between intermittent reinforcements of pennies (Kachanoff, Leveille, McLelland & Wayner, 1973). It may be that SIP reduces the arousal produced by intermittent reinforcement. Support for this view comes from the finding that SIP reduced pituitary-adrenal activity which was elevated in situations of high arousal (Brett & Levine, 1979). A hypothesis which has received considerable experimental attention is that SIP is due to a dry mouth. Injection of water into the mouth and ingestion of some liquid diets can abolish SIP. Kissileff (1973) argues that the dry mouth explanation is inadequate because SIP differs in a number of ways from the food-related drinking of rats with a persistently dry mouth following removal of the salivary glands. Also, if the interval between food pellets was long, drinking was delayed, indicating that the drinking was not simply due to a dry mouth (Segal, Oden & Deadwyler, 1965). Various other theories for SIP have been put forward, for example that it is caused by emotional induction, that it is simply a time filler between food reinforcements, or that the food pellets reinforce the drinking.

Effects of flavour on drinking

In Chapter 5 we saw that offering a variety of tastes or smells increases fluid intake, and that simply adding a non-nutritive sweetener also causes polydipsia. Rats which are in fluid balance will drink large volumes of saccharin solution and will go transiently into positive fluid balance. This is thus an example of drinking which is not in response to body fluid deficits. The excess water ingested is, however, rapidly excreted in a dilute urine. As with SIP, homeostatic control of the body fluids can be upset by impairing renal function. Rats drinking saccharin after an injection of antidiuretic hormone went markedly into positive fluid balance (i.e. by 22 mosmoles/kg H_2O). As the rats went into positive fluid balance the rate of drinking declined, but drinking did not stop (figure 5.6). The inhibitory signals from plasma dilution must be relatively ineffective against the consumption of a palatable saccharin solution. Taste overrides the homeostatic control of fluid balance (B. J. Rolls, Wood & Stevens, 1978). It seems that strong inhibitory thirst signals from plasma dilution may not have evolved because normally the kidneys cope adequately with excess fluid intake. This experiment stresses the importance of both fluid intake and output in body fluid homeostasis.

Not only can the palatability of fluid be increased, it can also be decreased by adding the bitter tasting substance quinine. When a quinine solution is the only source of fluid, rats do not always respond to body fluid deficits in the same way as they would if drinking water. After intraperitoneal or intravenous administration of hypertonic saline, rats drank quinine only after a long delay. When they did drink they took about half the amount needed to dilute the load to isotonicity. It is possible that this delay in drinking was due to the stress which accompanied acute administration of hypertonic saline. Less stressful procedures, such as fluid deprivation, heat dehydration, or extracellular depletion with colloid, all stimulated rats to drink about half as much quinine as water (Rowland, 1977). Intracranial injection of angiotensin also stimulated quinine drinking and the amount consumed depended on the concentration of the quinine (B. J. Rolls, Jones & Fallows, 1972; see figure 1.5). Clearly, taste can influence the amount consumed in response to fluid deficits, but this does not necessarily mean that the behaviour is non-regulatory. There are no studies which indicate that rats drinking quinine solution had dangerous dehydration or hypovolaemia. Indeed, when rats were offered quinine solution chronically as the only fluid source, although they were hypodipsic and plasma sodium and osmolality were somewhat elevated, the animals did not show signs of stress or ill health (Nicolaïdis & Rowland, 1975a). When rats, and presumably also other animals, are forced to meet fluid requirements by consuming noxious substances they appear to learn quickly that no untoward effects follow a reduction in fluid intake. This is because of the ability of the kidney to respond to reduced intake by the production of a concentrated urine and because the body fluids need to be maintained within a safe range, not at a precisely fixed point.

Thus, studies of flavour changes illustrate the flexibility of drinking behaviour. Animals appear to learn quickly about the consequences of voluntarily increasing or decreasing fluid intake. These studies, rather than showing that drinking is non-regulatory as is often claimed, stress that terms such as 'regulation' and 'homeostasis' can only be understood in the context of the animal's environment and the ability of the kidney to maintain fluid balance in the face of these environmental demands. In animals with normally functioning brains and kidneys and at least limited access to some fluid source, there are no examples of dangerous imbalances in the body fluids due to so-called non-regulatory drinking, although clearly the effects of flavour on drinking provide examples of how drinking can be influenced by factors other than body fluid deficits.

Other oropharyngeal factors, including a dry mouth, and temperature

During drinking it is not just the taste of the fluid which is important. A number of other factors, such as the motor act of drinking, the tactile and thermal properties of the fluid and a dry mouth (see Chapter 1), may all influence behaviour. The role of oropharyngeal factors in thirst is discussed in Chapter 5. However, the effects of water temperature on intake warrant further discussion when considering whether drinking is always regulatory. In rats on a once-a-day drinking schedule, cool water was much more effective in suppressing intake than warm water (see figure 10.2). This effect persisted in rats drinking water of different temperatures over the entire 24 h. Detailed body fluid analysis during and immediately after consumption of water of different temperatures is not yet available. It is quite likely that the water temperature affects the distribution of fluid. For example, cool water leaves the stomach more rapidly than warm fluid, whereas warm water is absorbed more rapidly from the gut into the bloodstream. The gut is apparently involved in the effects of temperature on drinking in that vagotomy abolished the effect (see Gold & Laforge, 1977). It is possible that, when body fluid analysis is available, it will show that at least part of the effect of temperature on drinking may be due to altered body fluid status.

Learning and drinking

Animals must learn to take advantage of the environment. They must learn to avoid dangerous food and drink and to subsist on limited resources. A variety of species has been shown to acquire an aversion

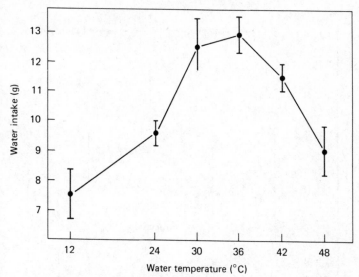

Fig. 10.2. Rat's mean water intake as a function of water temperature ($N = 8$). Water was available for 30 min/day. (Modified from Kapatos & Gold, 1972a.)

for flavoured fluids (the conditioned stimuli) which are followed by feelings of illness (the unconditioned stimulus) even when the delay between tasting the fluid and the feeling of illness is prolonged for up to 1 h. This conditioned taste aversion or bait-shyness has obvious adaptive value especially for omnivores, such as the rat, which subsist on a wide variety of food and drink. Classical conditioning of taste aversion forms an important part of learning theory and has been discussed widely in that context. In the context of a discussion of the controls of drinking such studies further emphasize the importance of taste for fluid intake. Animals can learn to associate particular tastes with the post-ingestive state and such learning will affect subsequent encounters with familiar fluids (see Barker, Best & Domjan, 1977).

Not only do animals appear to be able to learn which substances to drink, they can also learn to drink in particular situations. In typical classical conditioning paradigms previously neutral stimuli have been shown to elicit drinking following their repeated association with thirst-inducing treatments. These treatments include hypertonic NaCl, intra-cranial angiotensin injections, water deprivation, and hypovolaemia caused by subcutaneous formalin (see Weisinger, 1975). An example of such conditioned drinking is that if formalin injections were repeatedly paired with the odour of menthol (the conditioned stimulus), eventually

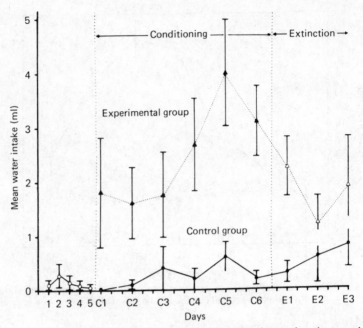

Fig. 10.3. Conditioning of water intake after hypovolaemia caused by formalin injections (unconditioned stimulus) to the odour of menthol (conditioned stimulus). Open circles represent all rats on preconditioning trials; solid circles and open triangles represent rats receiving control injections on that trial; solid triangles represent rats receiving formalin injections on that trial. The elevated intake during the first extinction trial suggests that conditioning has occurred. (From Weisinger, 1975.)

rats drank when they smelled the menthol even though there had been no formalin injection and there were no measurable plasma changes. Typical of such conditioning, this drinking rapidly extinguished when the menthol was presented without the formalin (see figure 10.3). Although more work is needed in this field to understand, for example, the types of stimuli that can trigger drinking, it seems that animals can learn to associate some environmental cues with the urge to drink. More work is also needed to understand the functional significance of such conditioned drinking in an animal's fluid economy, although it is possible that the anticipatory drinking described in the next section is an example of such conditioning.

The interaction of feeding and drinking – anticipatory drinking

In most animals there is a close association between feeding and drinking. We will consider two possible reasons for this association:

first, that it is related to actual physiological states and needs; second, that there is learning involved, for example that animals learn to associate particular smells and tastes of foods with particular fluid requirements and learn to drink in anticipation of fluid deficits.

If animals are thirsty they tend to eat less. This is the appropriate physiological response since food would tend to increase an animal's total body sodium. In the rat thirst stimuli such as hypertonic saline, angiotensin, and water deprivation reduce the intake of a dry diet (see B. J. Rolls, 1975). Also, infusions of water which reduce plasma osmolality lead to increased feeding which would help to restore plasma concentration to normal (Deaux & Kakolewski, 1971). There is some evidence that the interaction between hunger and thirst takes place in the central nervous system (B. J. Rolls & McFarland, 1973).

The relation between feeding and drinking is not so clear if animals are deprived of food. This should result in a decrease of fluid intake if the two are simply linked behaviourally. However, the internal changes associated with food deprivation are complex (Toates, 1979) and species vary widely in their fluid intake during food deprivation. More detailed body fluid analysis is required to understand the complex changes in fluid balance which occur in this situation.

As feeding and drinking are closely associated it is possible that much spontaneous drinking might be accounted for by physiological changes associated with feeding. Deaux, Sato & Kakolewski (1970) found that, in rats fed discrete meals, drinking was preceded by a rise in osmolality. A similar picture was found in dogs (see next section). Fitzsimons (1972) argues that such meal-fed rats may behave differently to those with free access to food. Meal-fed rats completed the meal before drinking whereas rats with free access to food drank as much as 30 % of the water associated with a meal before the meal had begun. He argues that, rather than drinking being a consequence of the physiological changes following feeding, it could be that the taste or smell of food or its physical presence in the mouth or stomach could cause drinking. Each food is associated with a particular metabolic need for water. Animals may be able to associate the particular characteristics of a food with previously learned fluid requirements and after several encounters antici- pate fluid needs. Fitzsimons & Le Magnen (1969) determined the effect of changing from a food associated with a low fluid requirement (i.e. high in carbohydrate) to one with a high fluid requirement (i.e. high in protein) (figure 10.4). The rats increased fluid intake during the first day after the diets were changed, but for the first 3 days after the change much of the drinking occurred some time after feeding. This delayed drinking could be due to systemic changes resulting from the

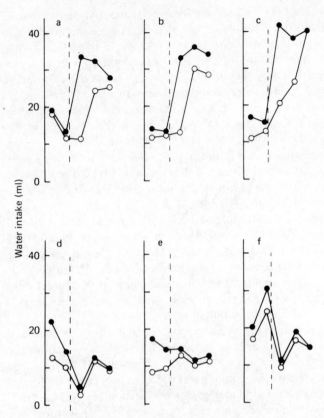

Fig. 10.4. Total daily water intake (solid circles) and intake of water associated with meals (open circles) during (a, b, c) the transition from a carbohydrate diet (i.e. with a low fluid requirement) to a protein diet (i.e. with a high fluid requirement); and (d, e, f) the transition from a protein to a carbohydrate diet. Each graph is from one rat. The diet was changed at the dotted line. It appears that total daily water intake is increased in anticipation of the dietary requirement for water. (From Fitzsimons & Le Magnen, 1969.)

high protein intake. However, after 3 days most of the drinking occurred in relation to feeding, i.e. before physiological changes were likely to have occurred. Thus, it appears that the rats had learned to associate the new diet with an increased need for water and therefore anticipated fluid needs. In the literature on thirst much weight has been given to this study, because of the implication that drinking may not be in response to fluid deficits but may anticipate them. However, since the body fluids were not analysed, it is not known whether the drinking really did precede systemic changes. Clearly, this important study needs to be repeated with body fluid analysis. Rowland (1979) has looked for anticipatory drinking in rats eating a salty diet (chow supplemented with 3% sodium chloride), but found that most of the drinking occurred between meals with no trend to increase meal-associated anticipatory drinking.

In conclusion, to the extent to which anticipatory drinking does occur, it provides an example of drinking which is not in response to sensed body fluid deficits.

One of the most satisfactory ways to determine whether spontaneous drinking occurs in response to fluid needs or in anticipation of them is to monitor continuously plasma changes and drinking and to examine the relation between them, as described on pp. 161–4.

The termination of drinking

As described in Chapter 6, factors other than dilution and plasma expansion make a contribution to the termination of drinking. These factors include in at least some species oropharyngeal metering, gastric distension, stimulation of the gut by water, and hepatic–portal mechanisms. These presystemic factors involved in the termination of drinking thus provide further examples of influences on drinking which are not due to systemic body fluid changes.

Is water deprivation-induced drinking in response to body fluid deficits?

In Chapter 4, it is shown (1) that 24 h water deprivation leads to significant changes in the body fluids, in particular to cellular dehydration and reduction of plasma volume, (2) that the change in cellular dehydration is greater than the threshold for the elicitation of drinking, and (3) that preloads of water to restore selectively cellular hydration reduce the drinking by 64–72% and preloads of saline to restore selectively extracellular fluid volume reduce drinking by 5–30% (depending on the species). Thus, in water deprivation-induced drinking the drinking is in response to body fluid changes.

Spontaneous drinking, with free access to water.

When organisms have free access to a variety of palatable fluids, as is often the case in man, it is possible that fluid deficits occur only rarely and that fluid intake is usually in excess of actual requirements and in anticipation of deficits. It should be pointed out, however, that actual plasma changes have not been measured in such a situation. That there

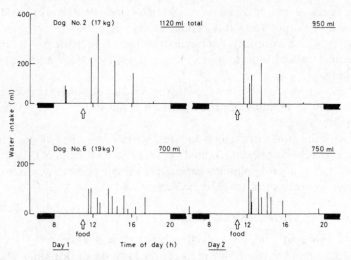

Fig. 10.5. Individual records of drinking by two dogs over a 48-h period when water was freely available in the home kennel. Food was given at 11.00 a.m. These dogs showed no anticipatory drinking; they drank in response to body fluid deficits. (From B. J. Rolls, Wood & Rolls, 1980.)

is a surfeit of fluid is often indicated by the copious dilute urine that man often produces. Man's abundant and varied fluid supply is probably not typical of many other species which would usually be confined to meeting fluid requirements from just water and food.

To study the mechanisms of spontaneous water drinking, we have analysed the body fluids of the dog to determine whether body fluid changes contribute to or account for drinking with *ad libitum* access to water. Typical patterns of daily water intake for two dogs are shown in figure 10.5. One feature of the drinking pattern of the dogs was that little water was drunk before and during feeding, but significant drinking usually started some time between 20 min and 60 min after the meal. It should be noted that these dogs were eating moist food (four parts food to one part water), so that they did not experience any difficulty swallowing.

Throughout the day the plasma was sampled every half hour, and immediately after every drinking bout. Plasma sodium levels during drinking and non-drinking periods are shown in table 10.1. The non-drinking plasma values were selected from samples taken during periods dissociated from drinking. It is shown in table 10.1 that plasma sodium was higher at the time of drinking (148.4 mequiv./l) than in non-drinking periods (144.2 mequiv./l). To determine whether this increase in cellular dehydration (equivalent to 4.2 mequiv./l of plasma sodium) was of

Table 10.1. *The relationship between spontaneous water intake and plasma sodium changes in the dog.*

	Feeding time	First drink after feeding	All non-drinking periods	All drinking bouts	Drinking after threshold dose of intravenous NaCl
Water intake (ml/kg)		8.0±1.1		7.6±0.8	2.5±0.1
Plasma sodium (m equiv./l)	144.3±1.4	149.3±1.1	144.2±1.3	148.4±1.1	149.5±0.7
Plasma sodium change (m equiv./l)		4.9±0.6		4.2±0.4	+3.6±0.4

Means are shown for water intake, plasma sodium concentration, and plasma sodium changes between conditions, for 11 dogs over 24-h periods with free access to water in the home kennel. The minimum change in plasma sodium concentration necessary to evoke drinking in ten of these dogs during intravenous hypertonic NaCl infusions is also shown when spontaneous drinking occurred, plasma sodium was elevated above the change necessary to evoke drinking. (From Rolls, Wood & Rolls, 1980.)

sufficient magnitude to initiate drinking, the plasma sodium change (produced by intravenous infusion of hypertonic sodium chloride) that was just necessary to initiate drinking in fluid-replete animals was also measured. This was 3.6 mequiv./l, so it can be concluded that the cellular dehydration present when the dog drinks under *ad libitum* conditions is of sufficient magnitude to induce drinking. These experiments provide evidence that in the dog *ad libitum* drinking may not be simply anticipatory, and that fluid deficits do occur under these conditions, can be sensed physiologically, and are large enough to produce drinking.

Pliny the Elder noted (A.D. 65–75) that water consumption is a function of the amount of salt in the diet. Certainly dietary salt is an important factor in the spontaneous drinking of our dogs. It is shown in table 10.1 that the effect of feeding the dogs was to elevate plasma sodium. At the time when substantial meal-related drinking first occurred, the increase in plasma sodium concentration (+4.9 mequiv./l) was greater than the change sufficient to evoke drinking. This indicates that the osmotic consequences of feeding in producing cellular dehydration may provide a physiological stimulus for drinking that is of importance in the natural environment. To test this further, the dogs were fed a low-sodium meal, to minimize the osmotic effects of feeding.

It was found that the latency to drink from the time of feeding was increased in the dogs on the low-sodium diet (from 48 ± 5 min to 244 ± 44 min, $N = 10$), and that the total daily volume of water ingested was decreased (from 42 ± 5 ml/kg to 21 ± 5 ml/kg).

It is interesting to note that changes in plasma volume may also be associated with drinking. In the sheep, feeding led to a rapid reduction in plasma volume and elevated plasma renin (Blair-West & Brook, 1969). The clearest plasma volume change in the dogs used in our study was at the time when drinking first occurred after feeding. Thus, plasma protein concentration increased from a mean value of 5.2 ± 0.1 to 5.5 ± 0.1 g% during the interval between feeding and drinking, equivalent to a change of about 6% in plasma volume. It has not been established whether this order of change in plasma volume is an effective thirst stimulus in the dog, but these observations further emphasize that the physiological changes which arise because of the ingestion of food are among the factors that underlie drinking under natural conditions.

The pattern of water intake which we have described for the dog appears to represent behaviour driven by significant physiological needs incurred as a result of the ingestion of food. The extent to which drinking in different species may be due to physiologically significant deficits in body fluid balance, or may be secondary or anticipatory, is of particular interest when considering physiological mechanisms and normal behaviour. Species differences in drinking apparent in the laboratory may reflect adaptation to environmental constraints during evolution, in particular in the nature of the diet and water resources. Future research, embracing studies of the natural drinking behaviour and ecology of animals, would be useful to further our understanding of the relation of physiological thirst mechanisms to normal behaviour. Suprisingly, studies of man in this regard are even less common than in other species.

Conclusions

In this chapter some of the environmental circumstances which have been found to affect drinking behaviour have been discussed. Much of the evidence cited has been used by others to argue that drinking is often non-regulatory or non-homeostatic. The concept of homeostasis should not be used to imply that an animal is like a simple thermostat that automatically switches on or off according to fixed bodily conditions. To assume that an animal always behaves in such an automatic fashion ignores the fact that it has a brain which makes it adaptable, able to benefit and learn from environmental change. It also ignores the fact that animals have efficient kidneys that can maintain the body fluids within safe limits when fluid and electrolyte intake fluctuate.

Because animals are adaptable there are bound to be experiments that show that the environment can influence drinking. Although the plasma concentration is usually maintained within relatively narrow limits when water is freely available, it can fluctuate more widely without untoward effects on the animal. It is therefore not surprising that animals with complex brains can learn that altering fluid intake in response to environmental circumstance rather than physiological need may not have dire consequences, and may sometimes be beneficial. This adaptability of drinking behaviour is not inconsistent with the concept of homeostasis, especially when homeostasis of the whole animal is considered.

More studies are needed to understand the way in which physiological and environmental factors interact to influence drinking. Experiments which combine an ethological approach to drinking with physiological measurements will aid our understanding of natural drinking behaviour. Examples of the kind of ethological studies which could be combined with body fluid analysis are as follows.

McFarland (1971) looked at the way in which water-deprived doves switched from feeding to drinking. The birds pecked one key for food and another for water and these two keys could be separated by a barrier of variable length. As the length of the barrier increased the birds persisted more with one activity before switching to the other. Thus, the energy cost of switching seems to have an influence on drinking. Another economic analysis of drinking has recently been reported by Marwine & Collier (1979). In the first study the rats were in a situation which could be the laboratory equivalent of having to travel a long way to gain access to water. The rats had to expend variable amounts of energy in pressing a bar merely to gain access to water. It was found that dramatic changes in drinking pattern occurred in this situation. Even when modest energy expenditure was required, the rats adopted the strategy of drinking only a few times in a 24-h period and taking large draughts when they did reach water. On the other hand, rats with free access to water, but which had to work for water at the consummatory stage by pressing the lever for a few drops of water, were relatively unaffected by the work load.

Summary

Some influences on drinking are 'non-homeostatic' or 'non-regulatory', in the sense that they are not due to disturbances of the body fluids. Examples of such influences on drinking are schedule-induced polydipsia, flavour-induced changes in drinking, oropharyngeal factors such as a dry mouth, the effect of water temperature on the amount consumed, the role of learning in affecting fluid intake, including anticipatory drinking, the maintenance of drinking (which does not depend, at least in the first instance, on changes in the body fluids, but rather on oropharyngeal stimulation) and the termination of drinking (which is partly controlled by presystemic factors such as gastric distension). Drinking induced by water deprivation is in many cases homeostatic, in response to body fluid deficits, as shown by measurements of deficit signals, which exceed the thirst threshold, and by selective deficit-replacement experiments (often using preloads). Spontaneous drinking with free access to water has been investigated less extensively, but in the dog is in response to significant plasma deficits. More studies which combine physiological, ethological and psychological techniques are needed to gain a more complete understanding of normal drinking behaviour.

11   Future directions

In this chapter we focus attention on a number of topics in which we believe that further developments are needed and are possible in the next few years. We hope that these thoughts will help to give a clear view of our present state of knowledge, and will stimulate our readers to produce their own ideas for experiments which will advance our understanding of brain and body function and behaviour, and of problems which arise clinically due to disturbed function.

Measurements of body fluid parameters and plasma variables

As stressed in this book, in order to assess the role of body fluid changes in drinking, and whether drinking is in this sense 'homeostatic' (see Chapter 10), analysis of body fluid parameters is needed. Cellular dehydration can be assessed by plasma sodium concentration and osmolality, and plasma volume by plasma protein concentration and haematocrit. In many experiments, for example when investigating the termination of drinking and the initiation of *ad libitum* drinking with free access to water, frequent sampling is needed. It is hoped that a reliable osmometer which requires only small samples will be developed in the near future (currently available models do not give accurate measurements with small samples) since this would make serial sampling in small animals such as the rat feasible, and should help advances to be made. The sampling could be through an indwelling catheter so that there is no disturbance. It is also useful to be able to measure levels of hormones such as renin, angiotensin and antidiuretic hormone in the plasma in order to determine what factors control different types of drinking.

Spontaneous drinking

Use of the above techniques will be necessary to determine to what extent normal drinking with free access to water is in response to particular body fluid changes and whether it is mediated by particular hormonal changes. Measured changes will then have to be compared with the threshold change for the elicitation of drinking. This comparison has been made for plasma sodium and osmolality in the dog, and it is clear that in this species, even with free access to water, the changes in plasma sodium exceed the threshold value for the elicita-

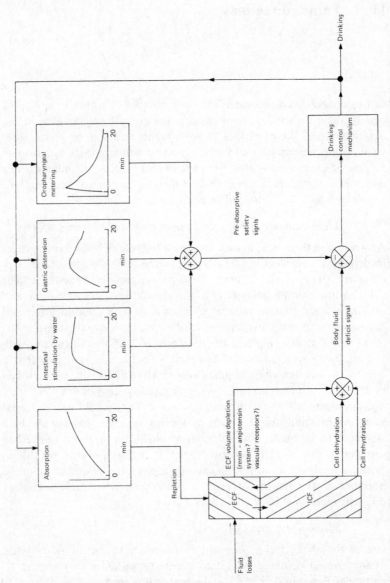

Fig. 11.1. Control model of drinking consistent with many of the findings described in this book. The input to the system, represented at the left of the diagram, consists of the fluid lost to the environment by evaporation, sweating, excretion, etc. A component of the fluid losses reduces the volume of the extracellular fluid compartment, and this depletion is sensed, perhaps by the vascular receptors or the renin–angiotensin system, to produce an extracellular fluid depletion signal. The osmotic component of

tion of drinking (see Chapter 10). Then the role of putative control signals can be further assessed by selectively restoring a disturbance in one such signal that is normally associated with drinking. For example, the effect on drinking of blocking or abolishing the effects of elevated plasma renin, or of intravenous infusions of water which prevent cellular dehydration (see Chapter 4), can be determined. These experiments should all be performed with and without food available, to determine to what extent body fluid changes following the digestion and absorption of food produce drinking.

Learning and strategy

Given that in at least some situations drinking is not in response to body fluid disturbances, and that through learning drinking can anticipate body fluid changes, it will be interesting to investigate conditioned appetite for water and conditioned satiety for water. The ways in which drinking is adapted to the environment, and the ways in which the pattern of drinking is altered when the difficulty of obtaining water is changed also comprise a developing subject (see Chapter 10).

Drinking in different species, including man

The fact that the pattern of drinking varies between species (see Chapter 6) provides an opportunity to analyse different controls of drinking. For example, pre-absorptive satiety mechanisms are likely to be particularly important and open to analysis in rapid drinkers such as the dog. Also,

> the fluid depletion (i.e. lost water) increases the concentration of the body fluids, results in shrinkage of the intracellular fluid compartment, and gives rise to a cellular fluid depletion signal. These two depletion signals are summed (represented by the first summation junction) to produce a signal which reflects the body fluid deficit, and acts through a drinking control mechanism to produce the output of the system, namely drinking, which is represented on the right of the diagram. The consequences of drinking lead to four signals which are activated in sequence and tend to stop drinking. These are shown in the loops at the top of the diagram. Of the first three, oropharyngeal metering has a rapid time-course of onset (indicated by the graph), and decays after drinking when the oropharynx is no longer stimulated by water. Gastric distension builds up over the first few minutes, and decays during a period of many minutes to hours. Intestinal stimulation by water also builds up gradually, and is maintained by slow gastric emptying for minutes or even hours. These three satiety signals are pre-absorptive, and because they can act before the body fluid loss is reversed, are short-term, temporary, satiety mechanisms. (They are shown as summing because this is a simple and stable assumption.) The last satiety mechanism shown is the repletion of the body-fluid deficits following absorption, and this is a long-term feedback mechanism, in that it reverses the original input to the system. (The authors found consultation with D. J. McFarland helpful in producing this diagram.)

work on the controls of drinking in man, and on clinical problems of thirst, is still at a relatively early stage, and has the advantage that subjective reports of thirst-related sensations can also be studied and the underlying factors analysed.

Mechanisms for drinking in response to extracellular thirst stimuli

Although great progress has been made in understanding renin–angiotensin-induced thirst (see Chapter 4), the relevance of this system to normal drinking is still not clear. Investigation of another system for drinking to extracellular stimuli, perhaps involving vascular distension receptors in the low-pressure circulation, is still at an early stage (see Chapter 4) and should be developed.

Models of drinking

With the identification of several different factors which can contribute to the initiation, maintenance and termination of drinking, a next stage is to determine quantitatively the role of each factor. This amounts to specifying a control system for drinking based on physiological measurements, in which each factor can be specified quantitatively. Models of drinking based on control theory have been specified in the past by McFarland (1965), Oatley (1967), Reeve & Kulhanek (1967), McFarland & McFarland (1968), Toates & Oatley (1970) and Toates (1980), and will continue to develop. In simulations these models allow quantitative testing of hypotheses as knowledge of the real system develops. A model which includes some of the factors described in this book is shown in figure 11.1.

The neural control of drinking

With progress in the identification of neural systems which contain sensors for cellular dehydration and for angiotensin-induced drinking it becomes possible to trace pathways neuroanatomically from these regions to other parts of the brain (see Chapter 7) and then to investigate how these brain regions are involved in the control of drinking. The method of analysing the role of such brain regions in drinking, using analysis of responses shown by single neurons during drinking, is still relatively new, and provides a useful method for tracing systems related to thirst through the brain and identifying the function being performed at each stage of the system. The pharmacology of drinking must eventually be understood in terms of how pharmacological agents influence the different elements in this system.

References

Abdelaal, A. E., Mercer, P. F. & Mogenson, G. J. 1976. Plasma angiotensin II levels and water intake following β-adrenergic stimulation, hypovolemia, cellular dehydration and water deprivation. *Pharmacology Biochemistry and Behavior*, **4**, 317–21.

Abraham, S. F., Baker, R. M., Blaine, E. H., Denton, D. A. & McKinley, M. J. 1975. Water drinking induced in sheep by angiotensin – a physiological or pharmacological effect? *Journal of Comparative and Physiological Psychology*, **88**, 503–18.

Abraham, S. F., Denton, D. A. & Weisinger, R. S. 1976. Effect of an angiotensin antagonist, Sar¹-Ala⁸-Angiotensin II on physiological thirst. *Pharmacology Biochemistry and Behavior*, **4**, 243–7.

Adachi, A., Niijima, A. & Jacobs, H. L. 1976. An hepatic osmoreceptor mechanism in the rat: electrophysiological and behavioral studies. *American Journal of Physiology*, **231**, 1043–9.

Adolph, E. F. 1939. Measurements of water drinking in dogs. *American Journal of Physiology*, **125**, 75–86.

Adolph, E. F. 1947. *Physiology of Man in the Desert*, New York: Interscience.

Adolph, E. F. 1950. Thirst and its inhibition in the stomach. *American Journal of Physiology*, **161**, 374–86.

Adolph, E. F. 1964. Opening Address. Regulation of body water content through water ingestion. In *Thirst*, ed. M. J. Wayner, pp. 5–14. Oxford: Pergamon.

Adolph, E. F., Barker, J. P. & Hoy, P. A. 1954. Multiple factors in thirst. *American Journal of Physiology*, **178**, 538–62.

Andersson, B. 1952. Polydipsia caused by intrahypothalamic injections of hypertonic NaCl solutions. *Experientia*, **8**, 157–8.

Andersson, B. 1953. The effect of injections of hypertonic NaCl solutions into different parts of the hypothalamus of goats. *Acta Physiologica Scandinavica*, **81**, 188–201.

Andersson, B. 1978. Regulation of water intake. *Physiological Reviews*, **58**, 582–603.

Andersson, B., Dallman, M. F. & Olsson, K. 1969. Observations on central control of drinking and on the release of antidiuretic hormone (ADH). *Life Sciences*, **8**, 425–32.

Andersson, B. & Eriksson, L. 1971. Conjoint action of sodium and angiotensin on brain mechanisms controlling water and salt balances. *Acta Physiologica Scandinavica*, **81**, 18–29.

Andersson, B. & McCann, S. M. 1955. A further study of polydipsia evoked by hypothalamic stimulation in the goat. *Acta Physiologica Scandinavica*, **33**, 333–46.

Andersson, B. & Olsson, K. 1973. On central control of body fluid homeostasis. *Conditional Reflex*, **8**, 147–59.

Andersson, B. & Westbye, O. 1970. Synergistic action of sodium and angiotensin on brain mechanisms controlling fluid balance. *Life Sciences*, **9**, 601–8.

Antelman, S. M., Rowland, N. E. & Fisher, A. E. 1976. Stimulation bound ingestive behavior: a view from the tail. *Physiology and Behavior*, **17**, 743–8.

Arieff, A. I., Guisado, R. & Lazarowitz, V. C. 1977. Pathophysiology of hyperosmolar states. In *Disturbances in Body Fluid Osmolality*, ed. T. E. Andreoli, J. J. Grantham and F. C. Rector, Jr, pp. 227–50. Bethesda, Md: American Physiological Society.

Arnauld, E., Dufy, B. & Vincent, J. D. 1975. Hypothalamic supraoptic neurones: rates and patterns of action potential firing during water deprivation in the unanaesthetized monkey. *Brain Research*, **100**, 315–25.

Avrith, D. B. & Fitzsimons, J. T. 1980. Increased sodium appetite in the rat induced by intracranial administration of components of the renin-angiotensin system. *Journal of Physiology (London)*, **301**, 349–64.

Barker, L. M., Best, M. R. & Domjan, M. (eds) 1977. *Learning Mechanisms in Food Selection*. Waco, Texas: Baylor University Press.

Bellows, R. T. 1939. Time factors in water drinking in dogs. *American Journal of Physiology*, **125**, 87–97.

Bernard, C. 1856. *Leçons de physiologie expérimentale appliquée à la médicine faites au Collège de France*, Vol. 2. Paris: Baillière.

Blair-West, J. R. & Brook, A. H. 1969. Circulatory changes and renin secretion in sheep in response to feeding. *Journal of Physiology (London)*, **204**, 15–30.

Blank, D. L. & Wayner, M. J. 1975. Lateral preoptic single unit activity: effects of various solutions. *Physiology and Behavior*, **15**, 723–30.

Blass, E. M. 1974. Evidence for basal forebrain thirst osmoreceptors in rat. *Brain Research*, **82**, 69–76.

Blass, E. M. & Epstein, A. N. 1971. A lateral preoptic osmosensitive zone for thirst in the rat. *Journal of Comparative and Physiological Psychology*, **76**, 378–94.

Blass, E. M. & Hall, W. G. 1974. Behavioral and physiological bases of drinking inhibition in water deprived rats. *Nature*, **249**, 485–6.

Blass, E. M. & Hall, W. G. 1976. Drinking termination: interactions among hydrational, orogastric, and behavioral controls in rats. *Psychological Review*, **83**, 356–74.

Blass, E. M., Hall, W. G. & Teicher, M. H. 1979. The ontogeny of suckling and ingestive behaviors. In *Progress in Psychobiology and Physiological Psychology*, Vol. 8, ed. J. M. Sprague and A. N. Epstein, pp. 243–99. New York: Academic.

Blass, E. M. & Hanson, D. G. 1970. Primary hyperdipsia in the rat following septal lesions. *Journal of Comparative and Physiological Psychology*, **70**, 87–93.

Blass, E. M., Jobaris, R. & Hall, W. G. 1976. Oropharyngeal control of drinking in rats. *Journal of Comparative and Physiological Psychology*, **90**, 909–16.

Blass, E. M., Nussbaum, A. I. & Hanson, D. G. 1974. Septal hyperdipsia: specific enhancement of drinking to angiotensin in rats. *Journal of Comparative and Physiological Psychology*, **87**, 422–39.

Block, M. L., Vallier, G. H. & Glickman, S. E. 1974. Elicitation of water ingestion in the Mongolian gerbil (*Meriones unguiculatus*) by intracranial injections of angiotensin II and L-norepinephrine. *Pharmacology Biochemistry and Behavior*, **2**, 235–242.

Bolles, R. C. 1979. Toy rats and real rats: nonhomeostatic plasticity in drinking. *Behavioral and Brain Sciences*, **2**, 103.

Booth, D. A. 1968. Mechanism of action of norepinephrine in eliciting an eating response on injection into the rat hypothalamus. *Journal of Pharmacology and Experimental Therapeutics*, **160**, 336–48.

Bott, E., Denton, D. A. & Weller, S. 1965. Water drinking in sheep with oesophageal fistulae. *Journal of Physiology (London)*, **176**, 323–36.

Brett, L. P. & Levine, S. 1979. Schedule-induced polydipsia suppresses pituitary-adrenal activity in rats. *Journal of Comparative and Physiological Psychology*, **93**, 946–56.

Brimley, C. C. & Mogenson, G. J. 1979. Oral motor deficits following lesions of the central nervous system in the rat. *American Journal of Physiology*, **237**, R126–31.

Brosnihan, K. B., Berti, G. A. & Ferrario, C. M. 1979. Hemodynamics of central infusion of angiotensin II in normal and sodium-depleted dogs. *American Journal of Physiology*, **237**, H139–45.

Brown, B. & Grossman, S. P. 1980. Evidence that nerve cell bodies in the zona incerta influence ingestive behavior. *Brain Research Bulletin*, **5**, 593–7.

Brown, J. J., Curtis, J. R., Lever, A. F., Robertson, J. I. S., DeWardener, H. E. & Wing, A. J. 1969. Plasma renin concentration and the control of blood pressure in patients on maintenance haemodialysis. *Nephron*, **6**, 329–49.

Bruno, J. P. & Hall, W. G. 1980. The ontgeny of drinking behavior in rats. In *The Proceedings of the Seventh International Conference on the Physiology of Food and Fluid Intake*. Warsaw.

Bryant, R. W., Epstein, A. N., Fitzsimons, J. T. & Fluharty, S. J. 1980. Arousal of a specific and persistent sodium appetite in the rat with continuous intracerebroventricular infusion of angiotensin II. *Journal of Physiology (London)*, **301**, 365–82.

Buggy, J. 1978. Block of cholinergic-induced thirst after obstruction of anterior ventral third ventricle or periventricular preoptic ablation. *Society for Neuroscience Abstracts*, **4**, 172.

Buggy, J., Fink, G. D., Johnson, A. K. & Brody, M. J. 1977. Prevention of the development of renal hypertension by anteroventral third ventricular tissue lesions. *Circulation Research*, Supplement I, **40**, 1–110–I–117.

Buggy, J. & Fisher, A. E. 1976. Anteroventral third ventricle site of action for angiotensin-induced thirst. *Pharmacology Biochemistry and Behavior*, **4**, 657–60.

Buggy, J., Hoffman, W. E., Phillips, M. I., Fisher, A. E. & Johnson, A. K. 1979. Osmosensitivity of rat third ventricle and interactions with angiotensin. *American Journal of Physiology*, **236**, R75–82.

Buggy, J. & Johnson, A. 1977a. Anteroventral third ventricle periventricular ablation: temporary adipsia and persisting thirst deficits. *Neuroscience Letters*, **5**, 177–82.

Buggy, J. & Johnson, A. K. 1977b. Preoptic-hypothalamic periventricular lesions: thirst deficits and hypernatremia. *American Journal of Physiology*, **23**, R44–52.

Buranarugsa, P. & Hubbard, J. I. (1979). The neural organization of the rat subfornical organ in vitro and a test of the osmo- and morphine-receptor hypothesis. *Journal of Physiology (London)*, **291**, 101–16.

Burton, M. J., Rolls, E. T. & Mora, F. 1976. Effects of hunger on the responses of neurons in the lateral hypothalamus to the sight and taste of food. *Experimental Neurology*, **51**, 668–77.

Cabanac, M. 1971. Physiological role of pleasure. *Science*, **173**, 1103–7.

Cannon, W. B. 1919, The physiological basis of thirst. *Proceedings of the Royal Society, London*, **90**, 283–301.

Cannon, W. B. 1947. *The Wisdom of the Body*. London: Kegan Paul, Trench, Trubner & Co.

Carlson, N. R. 1977. Physiology of Behavior. Boston: Allyn & Bacon.

Chwalbinska-Moneta, J. 1979. Role of hepatic portal osmoreceptor in the control of ADH release. *American Journal of Physiology*, **236**, E603–9.

Chiaraviglio, E. 1979. Drinking behaviour in rats treated with isoprenaline, angiotensin II or angiotensin antagonists. *Journal of Physiology (London)*, **296**, 193–202.

Coburn, P. C. & Stricker, E. M. 1978. Osmoregulatory thirst in rats after lateral preoptic lesions. *Journal of Comparative and Physiological Psychology*, **92**, 350–61.

Cooling, M. J. & Day, M. D. 1974. Inhibition of renin–angiotensin induced drinking in the cat by enzyme inhibitors and by analogue antagonists of angiotensin II. *Clinical and Experimental Pharmacology and Physiology*, **1**, 389–96.

Cross, B. A. & Green, J. D. 1959. Activity of single neurones in the hypothalamus: effects of osmotic and other stimuli. *Journal of Physiology (London)*, **148**, 554–69.

Davies, D. P. 1973. Plasma osmolality and feeding practices of healthy infants in first three months of life. *British Medical Journal*, **2**, 340–2.

Davis, J. O. 1967. The physiology of congestive heart failure. In *Handbook of Physiology*, ed. W. F. Hamilton and P. Dow, section 2, *Circulation*, Vol. 3, pp. 2071–122. Washington, D.C.: American Physiological Society.

Day, R. P. & Reid, I. A. 1976. Renin activity in dog brain: enzymological similarity to Cathepsin D. *Endocrinology*, **99**, 93–100.

Deaux, E. & Kakolewski, J. W. 1971. Character of osmotic changes resulting in the initiation of eating. *Journal of Comparative and Physiological Psychology*, **74**, 248–53.

Deaux, E., Sato, E. & Kakolewski, J. W. 1970. Emergence of systemic cues evoking food-associated drinking. *Physiology and Behavior*, **5**, 1177–9.

De Caro, G., Mariotti, M., Massi, M. & Micossi, L. G. 1980. Dipsogenic effect of angiotensin II, bombesin and tachykinins in the duck. *Pharmacology Biochemistry and Behavior*, **13**, 229–33.

Desor, J. A., Maller, O. & Andrews, K. 1975. Ingestive responses of human newborns to salty, sour, and bitter stimuli. *Journal of Comparative and Physiological Psychology*, **89**, 966–70.

Deutsch, J. A. 1979. Intragastric infusion and pressure. *Behavioral and Brain Sciences*, **2**, 105.

Divac, I. 1975. Magnocellular nuclei of the basal forebrain project to the neocortex, brain stem and olfactory bulb. Review of some functional correlates. *Brain Research*, **93**, 385–98.

Dubovsky, S. L., Grabon, S., Berl, T. & Schrier, R. W. 1973. Syndrome of inappropriate secretion of antidiuretic hormone with exacerbated psychosis. *Annals of Internal Medicine*, **79**, 551–4.

Elfont, R. M., Epstein, A. N. & Findlay, A. L. R. 1980. The role of the subfornical organ in angiotensin-induced drinking in the North American Opossum. *Journal of Physiology (London)*, **301**, 49P.

Emmers, R. 1973. Interaction of neural systems which control body water. *Brain Research*, **49**, 323–47.

Eng, R., Miselis, R. R. & Salanga, G. 1980. Knife cuts to the anterior stalk of the subfornical organ produce drinking deficits to angiotensin II but not other dipsogenic challenges. *Abstracts for the Society of Neuroscience*, **6**, 33.

Epstein, A. N. 1960. Water intake without the act of drinking. *Science*, **131**, 497–8.

Epstein, A. N. 1967. Oropharyngeal factors in feeding and drinking. In *Handbook of Physiology*, section 6: *Alimentary Canal*, Vol. I, *Control of Food and Water Intake*, Ch. 15, pp. 197–218. Washington D.C.: American Physiological Society.

Epstein, A. N. 1971. The lateral hypothalamic syndrome: its implications for the physiological psychology of hunger and thirst. In *Progress in Physiological Psychology*, Vol. 4, ed. E. Stellar and J. M. Sprague, pp. 263–317. New York: Academic.

Epstein, A. N. 1976. Feeding and drinking in suckling rats. In *Hunger: Basic Mechanisms and Clinical Implications*, ed. D. Novin, W. Wyrwicka and G. Bray, pp. 193–202. New York: Raven Press.

Epstein, A. N. 1978. Consensus, controversies and curiosities. *Federation Proceedings*, **37**, 2711–6.

Epstein, A. N., Fitzsimons, J. T. & Rolls, B. J. 1970. Drinking induced by injection of angiotensin into the brain of the rat. *Journal of Physiology (London)*, **210**, 457–74.

Epstein, A. N., Fitzsimons, J. T. & Simons, B. J. 1969. Drinking caused by the intracranial injection of angiotensin into the rat. *Journal of Physiology (London)*, **200**, 98–100P.

Epstein, A. N. & Hsaio, S. 1975. Angiotensin as dipsogen. In *Control Mechanisms of Drinking*, ed. G. Peters, J. T. Fitzsimons and L. Peters-Haefeli, pp. 108–116. Berlin: Springer-Verlag.

Epstein, A. N., Spector, D., Samman, A. & Goldblum, C. 1964. Exaggerated prandial drinking in rats without salivary glands. *Nature*, **201**, 1324–3.

Epstein, A. N. & Teitelbaum, P. 1964. Severe and persistent deficits in thirst produced by lateral hypothalamic damage. In *Thirst*, ed. M. J. Wayner, pp. 395–406. Oxford: Pergamon.

Ernits, T. & Corbit, J. D. 1973. Taste as a dipsogenic stimulus. *Journal of Comparative and Physiological Psychology*, **83**, 27–31.

Evered, M. D. & Fitzsimons, J. T. 1976. Drinking induced by angiotensin in the pigeon (*Columbia livia*). *Journal of Physiology (London)*, **263**, 193–4P.

Evered, M. D. & Mogenson, G. J. 1976. Regulatory and secondary water intake in rats with lesions of the zona incerta. *American Journal of Physiology*, **230**, 1049–57.

Evered, M. D. & Mogenson, G. J. 1977. Impairment in fluid ingestion in rats with lesions of the zona incerta. *American Journal of Physiology*, **233**, R53–8.

Falk, J. L. 1961. Production of polydipsia in normal rats by an intermittent food schedule. *Science*, **133**, 195–6.

Falk, J. L. 1966. The motivational properties of schedule-induced polydipsia. *Journal of Experimental Analysis of Behavior*, **9**, 19–25.

Falk, J. L. 1977. The origin and functions of adjunctive behavior. *Animal Learning and Behavior*, **5**, 325–35.

Fanestil, D. D. 1977. Hyposmolar syndromes. In *Disturbances in Body Fluid Osmolality*, ed. T. E. Andreoli, J. J. Grantham and F. C. Rector, Jr, pp. 267–84. Bethesda, Md: American Physiological Society.

Felix, D. & Akert, K. 1974. The effect of angiotensin II on neurones of the cat subfornical organ. *Brain Research*, **76**, 350–3.

Felix, D. & Schlegel, W. 1978. Angiotensin receptive neurones in the subfornical organ. Structure–activity relations. *Brain Research*, **149**, 107–16.

Findlay, A. L. R., Elfont, R. M. & Epstein, A. N., 1980. The site of the dipsogenic action of angiotensin II in the North American opossum. *Brain Research*, **198**, 85–94.

Fisher, A. E. & Coury, J. N. 1962. Cholinergic tracing of a central neural circuit underlying the thirst drive. *Science*, **138**, 691–3.

Fisher, A. E. & Coury, J. N. 1964. Chemical tracings of neural pathways mediating the thirst drive. In *Thirst*, ed. M. J. Wayner, pp. 515–29. Oxford: Pergamon.

Fitzsimons, J. T. 1961a. Drinking by nephrectomized rats injected with various substances. *Journal of Physiology (London)*, **155**, 563–79.

Fitzsimons, J. T. 1961b. Drinking by rats depleted of body fluid without increase in osmotic pressure. *Journal of Physiology (London)*, **159**, 297–309.

Fitzsimons, J. T. 1964. Drinking caused by constriction of the inferior vena cava in the rat. *Nature*, **204**, 479–80.

Fitzsimons, J. T. 1966. The hypothalamus and drinking. *British Medical Bulletin*, **22**, 232–37.

Fitzsimons, J. T. 1969. The role of renal thirst factor in drinking induced by extracellular stimuli. *Journal of Physiology (London)* **201**, 349–68.

Fitzsimons, J. T. 1971. The physiology of thirst: a review of the extraneural aspects of the mechanisms of drinking. In *Progress in Physiological Psychology*, ed. E. Stellar and J. M. Sprague, Vol. 4, pp. 119–201, New York: Academic.

Fitzsimons, J. T. 1972. Thirst. *Physiological Review*, **52**, 468–561.

Fitzsimons, J. T. 1973. Historical perspectives in the physiology of thirst. In *The Neuropsychology of Thirst: New Findings and Advances in Concepts*, ed. A. N. Epstein, H. R. Kissileff and E. Stellar, pp. 3–33. New York: John Wiley & Sons.

Fitzsimons, J. T. 1976. The physiological basis of thirst. *Kidney International*, **10**, 3–11.

Fitzsimons, J. T. 1978. Angiotensin, thirst, and sodium appetite: retrospect and prospect. *Federation Proceedings*, **37**, 2669–75.

Fitzsimons, J. T. 1979. The physiology of thirst and sodium appetite. *Monographs of the Physiological Society*, no. 35. Cambridge: Cambridge University Press.

Fitzsimons, J. T. & Kaufman, S. 1977. Cellular and extracellular dehydration and angiotensin as stimuli to drinking in the common iguana *Iguana iguana*. *Journal of Physiology (London)*, **265**, 443–63.

Fitzsimons, J. T. & Kucharczyk, J. 1978. Drinking and haemodynamic changes induced in the dog by intracranial injection of components of the renin-angiotensin system. *Journal of Physiology (London)*, **276**, 419–34.

Fitzsimons, J. T., Kucharczyk, J. & Richards, G. 1978. Systemic angiotensin-induced drinking in the dog: a physiological phenomenon. *Journal of Physiology (London)*, **276**, 435–48.

Fitzsimons, J. T. & Le Magnen, J. 1969. Eating as a regulatory control of drinking. *Journal of Comparative and Physiological Psychology*, **67**, 273–83.

Fitzsimons, J. T. & Moore-Gillon, M. J. 1979. Short-latency, graded drinking in response to reduction in venous return in the dog. *Journal of Physiology (London)* **295**, 76P.

Fitzsimons, J. T. & Moore-Gillon, M. J. 1980a. Pulmo-atrial junctional receptors and the inhibition of drinking. *Journal of Physiology (London)* **307**, 74–5P.

Fitzsimons, J. T. & Moore-Gillon, M. J. 1980b. Drinking and antidiuresis in response to reductions in venous return in the dog: neural and endocrine mechanisms. *Journal of Physiology (London)* **308**, 403–16.

Fitzsimons, J. T. & Setler, P. E. 1975. The relative importance of central nervous catecholaminergic and cholinergic mechanisms in drinking in reponse to angiotensin and other thirst stimuli. *Journal of Physiology (London)*, **250**, 613–31.

Fitzsimons, J. T. & Simons, B. J. 1968. The effect of angiotensin on drinking in the rat. *Journal of Physiology (London)*, **196**, 39–41P.

Fitzsimons, J. T. & Simons, B. J. 1969. The effects on drinking in the rat of intravenous infusion of angiotensin, given alone or in combination with other stimuli of thirst. *Journal of Physiology (London)*, **203**, 45–57.

Forrest, J. N. & Singer, I. 1977. Drug-induced interference with action of antidiuretic hormone. In *Disturbances in Body Fluid Osmolality*, ed. T. E. Andreoli, J. J., Grantham and F. C. Rector, Jr, pp. 309–40. Bethesda, Md: American Physiological Society.

Freed, W. J. & Mendelson, J. 1977. Water-intake volume regulation in the rat: Schedule-induced drinking compared with water-deprivation-induced drinking. *Journal of Comparative and Physiological Psychology*, **91**, 564–73.

Freis, E. D. 1976. Salt, volume and the prevention of hypertension. *Circulation*, **53**, 589–95.

Gallistel, C. R. & Beagley, G. 1971. Specificity of brain-stimulation reward in the rat. *Journal of Comparative and Physiological Psychology*, **76**, 199–205.

Gamble, J. L. 1954. *Chemical Anatomy, Physiology and Pathology of Extracellular Fluid*, 6th edn. Boston: Harvard University Press.

Ganong, W. F. 1979. *Review of Medical Physiology*, 9th edn. Los Altos, Cal.: Lange.

Garcia, J., Hankins, W. G. & Rusiniak, K. 1974. Behavioral regulation of the milieu interne in man and rat. *Science*, **185**, 824–31.

Gilman, A. 1937. The relation between blood osmotic pressure, fluid distribution and voluntary water intake. *American Journal of Physiology*, **120**, 323–8.

Gold, R. M., Kapatos, G., Prowse, J., Quackenbush, P. M. & Oxford, T. W. 1973. Role of water temperature in the regulation of water intake. *Journal of Comparative and Physiological Psychology*, **85**, 52–63.

Gold, R. M. & Laforge, R. G. 1977. Temperature of ingested fluids: preference and satiation effects (pease porridge warm, pease porridge cool). In *Drinking Behavior Oral Stimulation, Reinforcement, and Preference*, ed. J. A. W. M. Weijnen and J. Mendelson, pp. 247–74. New York: Plenum.

Goldstein, D. J. & Halperin, J. A. 1977. Mast cell histamine and cell dehydration thirst. *Nature*, **267**, 250–2.

Greer, M. A. 1955. Suggestive evidence of a primary 'drinking center' in hypothalamus of the rat. *Proceedings of the Society for Experimental Biology and Medicine*, **89**, 59–62.

Gregersen, M. I. & Cannon, W. B. 1932. Studies on the regulation of water intake. I. The effect of extirpation of the salivary glands on the water intake of dogs while panting. *American Journal of Physiology*, **102**, 336–43.

Grossman, S. P. 1960. Eating or drinking in satiated rats elicited by adrenergic or cholinergic stimulation, respectively, of the lateral hypothalamus. *Science*, **132**, 301–2.

Grossman, S. P. 1962a. Direct adrenergic and cholinergic stimulation of hypothalamic mechanisms. *American Journal of Physiology*, **202**, 872–82.

Grossman, S. P. 1962b. Effects of adrenergic and cholinergic blocking agents on hypothalamic mechanisms. *American Journal of Physiology*, **202**, 1230–6.

Grossman, S. P. 1967. *A Textbook of Physiological Psychology*. New York: John Wiley & Sons.

Grossman, S. P., Dacey, D., Halaris, A. E., Collier, T. & Routtenberg, A. 1978.

Aphagia and adipsia after preferential destruction of nerve cell bodies in hypothalamus. *Science*, **202**, 537–9.

Haack, D., Homsy, E., Kohrs, G., Möhring, B., Oster, P. & Möhring, J. 1975. Studies on drinking–feeding interactions in rats with hereditary hypothalamic diabetes insipidus. In: *Control Mechanisms of Drinking*, ed. G. Peters, J.T. Fitzsimons and L. Peters-Haefeli, pp. 41–4. Berlin: Springer-Verlag.

Haberich, F. J. 1968. Osmoreception in the portal system. *Federation Proceedings*, **27**, 1137–41.

Hall, W. G. 1973. A remote stomach clamp to evaluate oral and gastric controls of drinking in the rat. *Physiology and Behavior*, **11**, 897–901.

Hall, W. G. & Blass, E. M. 1975. Orogastric, hydrational, and behavioral controls of drinking following water deprivation in rats. *Journal of Comparative and Physiological Psychology*, **89**, 939–54.

Hall, W. G. & Blass, E. M. 1977. Orogastric determinants of drinking in rats: interaction between absorptive and peripheral controls. *Journal of Comparative and Physiological Psychology*, **91**, 365–73.

Haller, A. von. 1764. Fames et sitis. In *Elementa physiologiae corporis humani*, Vol. 6, pp. 164–87. Berne: Sumptibus Societatis Typographicae.

Hamilton, L. W. 1976. *Basic Limbic System Anatomy of the Rat*. New York: Plenum.

Hatton, G. I. & Bennett, C. T. 1970. Satiation of thirst and termination of drinking: roles of plasma osmolality and absorption. *Physiology and Behavior*, **5**, 579–87.

Hayward, J. N. 1977. Functional and morphological aspects of hypothalamic neurons. *Physiological Review*, **57**, 574–658.

Hayward, J. N. & Vincent, J. D. 1970. Osmosensitive single neurones in the hypothalamus of unanaesthetized monkeys. *Journal of Physiology (London)*, **210**, 947–71.

Hedin, S. 1899. *Through Asia*. Vol. I, II. New York & London: Harper.

Henderson, I. W., McKeever, A. & Kenyon, C. J. 1979. Captopril (SQ 14225) depresses drinking and aldosterone in rats lacking vasopressin. *Nature*, **281**, 569–70.

Hennessy, J. W., Grossman, S. P. & Kanner, M. 1977. A study of the etiology of the hyperdipsia produced by coronal knife cuts in the posterior hypothalamus. *Physiology and Behavior*, **18**, 73–80.

Hirano, T., Takei, Y. & Kobayashi, H. 1978. Angiotensin and drinking in the eel and frog. In *Osmotic and Volume Regulation*, ed. C. Barker Jorgenson and E. Skadhauge pp. 123–8. Copenhagen: Munksgaard.

Hirose, S., Yokosawa, H. & Inagami, T. 1978. Immunochemical identification of renin in rat brain and distinction from acid proteases. *Nature*, **274**, 392–3.

Hoffman, W. E., Ganten, U., Phillips, M. I., Schmid, P. G., Schelling, P. & Ganten, D. 1978. Inhibition of drinking in water-deprived rats by combined central angiotensin II and cholinergic receptor blockade. *American Journal of Physiology*, **234**, F41–7.

Hoffman, W. E. & Phillips, M. I. 1976. The effect of subfornical organ lesions and ventricular blockade on drinking induced by angiotensin II. *Brain Research*, **108**, 59–73.

Hogan, P. A. & Woolsey, R. M. 1967. Polydipsia associated with occult hydrocephalus. *New England Journal of Medicine*, **277**, 639–40.

Holmes, J. H. 1960. The thirst mechanism and its relation to edema. In *Edema*:

Mechanisms and Management, ed. J. H. Moyer and M. Fuchs, pp. 95–102 Philadelphia: Saunders.

Hosutt, J. A., Rowland, N. & Stricker, E. M. 1978. Hypotension and thirst in rats after isoproterenol treatment. *Physiology and Behavior*, **21**, 593–8.

Hosutt, J. A., Rowland, N. & Stricker, E. M. 1981. Impaired drinking responses of rats with lesions of the subfornical organ. *Journal of Comparative and Physiological Psychology*, **95**, 104–13.

Houpt, K. A. & Epstein, A. N. 1971. The complete dependence of beta-adrenergic drinking on the renal dipsogen. *Physiology and Behavior*, **7**, 897–902.

Hsaio, S., Epstein, A. N. & Camardo, J. S. 1977. The dipsogenic potency of peripheral angiotensin II. *Hormones and Behavior*, **8**, 129–40.

Huang, Y. H. & Mogenson, G. J. 1972. Neural pathways mediating drinking and feeding in rats. *Experimental Neurology*, **37**, 269–86.

Hunt, J. N. 1956. Some properties of an alimentary osmoreceptor mechanism. *Journal of Physiology (London)*, **132**, 267–88.

Hunt, J. N. & Stubbs, D. F. 1975. The volume and energy content of meals as determinants of gastric emptying. *Journal of Physiology (London)*, **245**, 209–25.

Jenner, F. A. & Eastwood, P. R. 1978. Renal effects of lithium. In *Lithium in Medical Practice*, ed. F. N. Johnson and S. Johnson, pp. 247–63. Lancaster: MTP Press.

Jewell, P. A. & Verney, E. B. 1953. Recent work on the localization of the osmoreceptors. *Journal of Endocrinology*, **9**, ii–iii.

Jewell, P. A. & Verney, E. B. 1957. An experimental attempt to determine the site of the neurohypophysial osmoreceptors in the dog. *Philosophical Transactions of the Royal Society*, **240B**, 197–324.

Johnson, A. K. & Buggy, J. 1978. Periventricular preoptic-hypothalamus is vital for thirst and normal water economy. *American Journal of Physiology*, **234**, R122–9.

Johnson, A. K. & Epstein, A. N. 1975. The cerebral ventricles as the avenue for the dipsogenic action of intracranial angiotensin. *Brain Research*, **86**, 399–418.

Johnson, A. K., Mann, J. F. E., Rascher, W., Johnson, J. K. & Ganten, D. 1981. Plasma angiotensin II concentrations and experimentally induced thirst. *American Journal of Physiology*, **240**, R229–34.

Jones, B. & Mishkin, M. 1972. Limbic lesions and the problem of stimulus–reinforcement associations. *Experimental Neurology*, **36**, 362–77.

Kachanoff, R., Leveille, R., McLelland, J. P. & Wayner, M. J. 1973. Schedule induced behavior in humans. *Physiology and Behavior*, **11**, 395–8.

Kapatos, G. & Gold, R. M. 1972a. Tongue cooling during drinking: A regulator of water intake in rats. *Science*, **176**, 685–6.

Kapatos, G. & Gold, R. M. 1972b. Rats drink less cool water: a change in the taste of water? *Science*, **178**, 1121.

Kare, M. R., Fregly, M. J. & Bernard, R. A. 1980. *Biological and Behavioral Aspects of Salt Intake*, New York: Academic.

Kenney, N. J. & Epstein, A. N. 1978. Antidipsogenic role of the E-prostaglandins. *Journal of Comparative and Physiological Psychology*. **92**, 204–19.

Kenny, J. T., Wright, J. W. & Reynolds, T. J. 1976. Schedule-induced polydipsia: the role of oral and plasma factors. *Physiology and Behavior*, **17**, 939–45.

Kievit, J. & Kuypers, H. G. J. M. 1973. Subcortical afferents to the frontal lobe in the rhesus monkey studied by means of retrograde horseradish peroxidase transport. *Brain Research*, **85**, 261–6.

Kissileff, H. R. 1973. Nonhomeostatic controls of drinking. In *The Neuropsychology of Thirst: New Findings and Advances in Concepts*, ed. A. N. Epstein, H. R. Kissileff and E. Stellar, pp. 163–98. Washington, D.C.: V. H. Winston.

Kissileff, H. R. & Epstein, A. N. 1969. Exaggerated prandial drinking in the 'recovered lateral' rat without saliva. *Journal of Comparative and Physiological Psychology*, **67**, 301–8.

Kourilsky, R. 1950. Diabetes insipidus. *Proceedings of the Royal Society of Medicine*, **43**, 842–4.

Kozłowski, S. & Drzewiecki, K. 1973. The role of osmoreception in portal circulation in control of water intake in dogs. *Acta Physiologica Polonica*, **24**, 325–30.

Kozłowski, S., Drzewiecki, K. & Zurawski, W. 1972. Relationship between osmotic reactivity of the thirst mechanism and the angiotensin and aldosterone level in the blood of dogs, *Acta Physiologica Polonica*, **23**, 417–25.

Kozłowski, S. & Szczepańska-Sadowska, E. 1975. Mechanisms of hypovolaemic thirst and interactions between hypovolaemia, hyperosmolality and the antidiuretic system. In *Control Mechanisms of Drinking*, ed. G. Peters, J. T. Fitzsimons and L. Peters-Haefeli, pp. 25–35. Berlin: Springer-Verlag.

Kraly, F. S. 1978. Abdominal vagotomy inhibits osmotically induced drinking in the rat. *Journal of Comparative and Physiological Psychology*, **92**, 999–1013.

Kraly, F. S., Gibbs, J. & Smith, G. P. 1975. Disordered drinking after abdominal vagotomy in rats. *Nature*, **258**, 226–8.

Kucharczyk, J., Assaf, S. Y. & Mogenson, G. J. 1976. Differential effects of brain lesions on thirst induced by the administration of angiotensin II to the preoptic region, subfornical organ and anterior third ventricle. *Brain Research*, **108**, 327–37.

Latta, T. 1832. Letter from Dr. Latta to the Secretary of the Central Board of Health, London, affording a view of the rationale and results of his practice in the treatment of cholera by aqueous and saline injections. *Lancet*, **ii**, 274–7.

Leaf, A. & Newburgh, L. H. 1955. *Significance of the Body Fluids in Clinical Medicine*, 2nd edn. Springfield, Ill.: Thomas.

Lee, M. C., Thrasher, T. N. & Ramsay, D. J. 1981. Is angiotensin essential in drinking induced by water deprivation and caval ligation? *American Journal of Physiology*, **240**, R75–80.

Leenen, F. H. H., De Jong, W. & De Wied, D. 1972. Water intake schedules and development of experimental hypertension in the rat. *Excerpta Medica International Congress Series*, No. 256, 587.

Leenen, F. H. H., Stricker, E. M., McDonald, R. H. & De Jong, W. 1975. Relationships between increase in plasma renin activity and drinking following different types of dipsogenic stimuli. In *Control Mechanisms of Drinking*, ed. G. Peters, J. T. Fitzsimons and L. Peters-Haefeli, pp. 84–8. Berlin: Springer-Verlag.

Lehr, D., Mallow, J. & Krukowski, M. 1967. Copious drinking and simultaneous inhibition of urine elicited by beta-adrenergic stimulation and contrary effect of alpha-adrenergic stimulation. *Journal of Pharmacology and Experimental Therapeutics*, **158**, 150–63.

Leibowitz, S. F. 1971. Hypothalamic alpha- and beta-adrenergic systems regulate both thirst and hunger in the rat. *Proceedings of the National Academy of Sciences, USA*, **68**, 440–4.

Leibowitz, S. F. 1975a. Pattern of drinking and feeding produced by hypothalamic norephinephrine injection in the satiated rat. *Physiology and Behavior,* **14,** 731–42.

Leibowitz, S. F. 1975b. Ingestion in the satiated rat: role of alpha and beta receptors in mediating effects of hypothalamic adrenergic stimulation. *Physiology and Behavior,* **14,** 745–54.

Leibowitz, S. F. 1981. Neurochemical systems of the hypothalamus in control of feeding and drinking behavior and water-electrolyte excretion. In *Handbook of the Hypothalamus,* ed. P. J. Morgane and J. Panksepp, Vol. 3, Part A, pp. 299–437. New York: Dekker.

Le Magnen, J. 1956. Hyperphagie provoquée chez le rat blanc par altération du mécanisme de satiété périphérique. *Comptes rendus de la Société de Biologie,* **150,** 32–5.

Lewis, P. R. & Shute, C. C. C. 1978. Cholinergic pathways in CNS. In *Handbook of Psychopharmacology,* Vol. 9, *Chemical pathways in the Brain,* ed. L. L. Iversen, S. D. Iversen and S. H. Snyder, Vol. 9, pp. 315–55. New York: Plenum.

Lind, R. W. & Johnson, A. K. 1980. Knife cuts between the subfornical organ (SFO) and antero-ventral third ventricle (AV3V) block drinking to peripheral angiotensin II. *Abstracts of the Society for Neuroscience,* **6,** 33.

Lindvall, O. & Björklund, A. 1978. Organization of catecholamine neurons in the rat central nervous system. In *Handbook of Psychopharmacology,* Vol. 9, *Chemical Pathways in the Brain,* ed. L. L. Iversen, S D. Iversen and S. H. Snyder, pp. 139–231. New York: Plenum.

Livett, B. G. 1973. Histochemical visualization of peripheral and central adrenergic neurones. *British Medical Bulletin,* **29,** 93–9.

Lotter, E. C., McKay, L. D., Mangiapane, M. L., Simpson, J. B., Vogel, K. E., Porte, D., Jr & Woods, S. C. 1980. Intraventricular angiotensin elicits drinking in the baboon. *Proceedings of the Society for Experimental Biology and Medicine,* **163,** 48–51.

Maddison, S., Rolls, B. J., Rolls, E. T. & Wood, R. J. 1977. Analysis of drinking in the chronically cannulated monkey. *Journal of Physiology (London),* **272,** 4–5P.

Maddison, S., Rolls, B. J., Rolls, E. T. & Wood, R. J. 1980. The role of gastric factors in drinking termination in the monkey. *Journal of Physiology (London),* **305,** 55–6P.

Maddison, S., Wood, R. J., Rolls, E. T., Rolls, B. J. & Gibbs, J. 1980. Drinking in the rhesus monkey: peripheral factors. *Journal of Comparative and Physiological Psychology,* **94,** 365–74.

Magendie, F. 1823. Histoire d'un hydrophobe, traité à l'Hôtel-Dieu de Paris, au moyen de l'injection de l'eau dans les veines. *Journal de Physiologie expérimentale et pathologique,* **3,** 382–92.

Malmo, R. B. & Mundl, W. J. 1975. Osmosensitive neurons in the rat's preoptic area: medial–lateral comparison. *Journal of Comparative and Physiological Psychology,* **88,** 161–75.

Malvin, R. L., Mouw, D. & Vander, A. J. 1977. Angiotensin: physiological role in water-deprivation-induced thirst in rats. *Science,* **197,** 171–3.

Malvin, R. L., Schiff, D. & Eiger, S. 1980. Angiotensin and drinking rates in the euryhaline killifish. *American Journal of Physiology,* **239,** R31–4.

Mangiapane, M. L. & Simpson, J. B. 1979. Pharmacologic independence of subfornical organ receptors mediating drinking. *Brain Research,* **178,** 507–17.

Mangiapane, M. L. & Simpson, J. B. 1980. Subfornical organ: forebrain site of pressor and dipsogenic action of angiotensin II. *American Journal of Physiology*, **239**, R382–9.

Mann, J. F. E., Schiffrin, E. L., Johnson, A. K., Boucher, R. & Genest, J. 1979. Modulation of the central effects of angiotensin II (A II) by sodium. *Society for Neuroscience Abstracts*, **5**, 221.

Mann, J. F. E., Johnson, A. K. & Ganten, D. 1980. Plasma angiotensin II: dipsogenic levels and angiotensin-generating capacity of renin. *American Journal of Physiology*, **238**, R372–7.

Marshall, J. F., Richardson, J. S. & Teitelbaum, P. 1974. Nigrostriatal bundle damage and the lateral hypothalamic syndrome. *Journal of Comparative and Physiological Psychology*, **87**, 808–30.

Marshall, J. F., Turner, B. H. & Teitelbaum, P. 1971. Sensory neglect produced by lateral hypothalamic damage. *Science*, **174**, 523–5.

Marwine, A. & Collier, G. 1979. The rat at the waterhole. *Journal of Comparative and Physiological Psychology*, **93**, 391–402.

Mayer, A. 1900. Variations de la tension osmotique de sang chez les animaux privés de liquides. *Comptes rendus des Séances de la Société de Biologie et de ses Filiales*, **52**, 153–5.

McCance, R. A. 1936. Experimental sodium chloride deficiency in man. *Proceedings of the Royal Society, London*, **119B**, 245–68.

McFarland, D. J. 1965. Control theory applied to the control of drinking in the Barbary dove. *Animal Behaviour*, **13**, 478–92.

McFarland, D. J. 1971. *Feedback Mechanisms in Animal Behaviour*. London: Academic.

McFarland, D. J. & McFarland, F. J. 1968. Dynamic analysis of an avian drinking response. *Medical and Biological Engineering*, **6**, 659–68.

McFarland, D. J. & Rolls, B. J. 1972. Suppression of feeding by intracranial injections of angiotensin. *Nature*, **236**, 172–3.

McKinley, M. J., Denton, D. A., Graham, W. F., Leksell, L. G., Mouw, D. R., Scoggins, B. A., Smith, M. H., Weisinger, R. S. & Wright, R. D. 1980. Lesions of the organum vasculosum of the lamina terminalis inhibit water drinking to hypertonicity in sheep. In *Proceedings of the Seventh International Conference on the Physiology of Food and Fluid Intake*. Warsaw.

McKinley, M. J., Denton, D. A., Leksell, L., Tarjan, E. & Weisinger, R. S. 1980. Evidence for cerebral sodium sensors involved in water drinking in sheep. *Physiology and Behavior*, **25**, 501–4.

Milgram, N. W. 1979. On the inadequacy of a homeostatic model: where do we go from here? *Behavioral and Brain Sciences*, **2**, 111–2.

Miller, N. E. 1965. Chemical coding of behavior in the brain. *Science*, **148**, 328–38.

Miller, N. E., Gottesman, K. S. & Emery, N. 1964. Dose response to carbachol and norepinephrine in rat hypothalamus. *American Journal of Physiology*, **206**, 1384–8.

Miller, N. E., Sampliner, R. I. & Woodrow, P. 1957. Thirst reducing effects of water by stomach fistula versus water by mouth, measured by both a consummatory and an instrumental response. *Journal of Comparative and Physiological Psychology*, **50**, 1–5.

Misantone, L. J., Ellis, S. & Epstein, A. N. 1980. Development of angiotensin-induced drinking in the rat. *Brain Research*, **186**, 195–202.

Miselis, R. R., Shapiro, R. E. & Hand, P. J. 1979. Subfornical organ efferents to neural systems for control of body water. *Science,* **205,** 1022–5.

Mogenson, C. J. & Kucharczyk, J. 1978. Central neural pathways for angiotensin-induced thirst. *Federation Proceedings,* **37,** 2683–8.

Möhring, J., Möhring, B., Haack, D., Lazar, J., Oster, P., Schömig, A. & Gross, F. 1975. Thirst and salt appetite in experimental renal hypertension of rats. In *Control Mechanisms of Drinking,* ed. G. Peters, J. T. Fitzsimons and L. Peters-Haefeli, pp. 155–62. Berlin: Springer-Verlag.

Montgomery, M. F. 1931. The role of the salivary glands in the thirst mechanism. *American Journal of Physiology,* **96,** 221–7.

Mook, D. G. 1963. Oral and postingestional determinants of the intake of various solutions in rats with oesophageal fistulas. *Journal of Comparative and Physiological Psychology,* **56,** 645–59.

Mora, F., Mogenson, G. J. & Rolls, E. T. 1977. Activity of neurones in the region of the substantia nigra during feeding. *Brain Research,* **133,** 267–76.

Morrison, S. D. 1968. The constancy of the energy expended by rats on spontaneous activity and the distribution of activity between feeding and non-feeding. *Journal of Physiology (London),* **197,** 305–23.

Mouw, D. R., Vander, A. J. & Wagner, J. 1978. Effects of prenatal and early postnatal sodium deprivation on subsequent adult thirst and salt preference in rats. *American Journal of Physiology,* **234,** F59–63.

Nachman, M. 1962. Taste preferences for sodium salts by adrenalectomized rats. *Journal of Comparative and Physiological Psychology,* **55,** 1124–9.

Nachman, M. & Ashe, J. H. 1974. Effects of basolateral amygdala lesions on neophobia, learned taste aversions, and sodium appetite in rats. *Journal of Comparative and Physiological Psychology,* **87,** 622–43.

Nicolaïdis, S. 1969. Early systemic responses to orogastric stimulation in the regulation of food and water balance: functional and electrophysiological data. *Annals of the New York Academy of Sciences,* **151,** 1176–203.

Nicolaïdis, S. & Rowland, N. 1974. Long-term self-intravenous 'drinking' in the rat. *Journal of Comparative and Physiological Psychology,* **87,** 1–15.

Nicolaïdis, S. & Rowland, N. 1975a. Regulatory drinking in rats with permanent access to a bitter fluid source. *Physiology and Behavior,* **14,** 819–24.

Nicolaïdis, S. & Rowland, N. 1975b. Systemic versus oral and gastrointestinal metering of fluid intake. In *Control Mechanisms of Drinking,* ed. G. Peters, J. T. Fitzsimons and L. Peters-Haefeli, pp. 14–21. New York: Springer.

Nothnagel, H. 1881. Durst und polydipsie. *Archiv für pathologische Anatomie und Physiologie,* **86,** 435–47.

Oatley, K. 1967. A control model of the physiological basis of thirst. *Medical and Biological Engineering,* **5,** 225–37.

Olds, J., Allan, W. S. & Briese, A. E. 1971. Differentiation of hypothalamic drive and reward centres. *American Journal of Physiology,* **221,** 368–75.

Oomura, Y., Ono, T., Ooyama, H. & Wayner, M. J. 1969. Glucose and osmosensitive neurones of the rat hypothalamus. *Nature,* **222,** 282–4.

Peck, J. W. & Blass, E. M. 1975. Localization of thirst and antidiuretic osmoreceptors by intracranial injections in rats. *American Journal of Physiology,* **228,** 1501–9.

Peck, J. W. & Novin, D. 1971. Evidence that osmoreceptors mediating drinking in rabbits are in the lateral preoptic area. *Journal of Comparative and Physiological Psychology,* **74,** 134–47.

Peterson, D. T. & Marshall, W. H. 1975. Polydipsia and inappropriate secretion of antidiuretic hormone associated with hydrocephalus. *Annals of Internal Medicine*, **83**, 675–6.

Phillips, M. I. 1978. Angiotensin in the brain. *Neuroendocrinology*, **25**, 354–77.

Phillips, M. I. & Felix, D. 1976. Specific angiotensin II receptive neurones in the cat subfornical organ. *Brain Research*, **109**, 531–40.

Phillips, M. I. & Hoffman, W. E. 1977. Sensitive sites in the brain for the blood pressure and drinking responses to angiotensin II. In *Central Actions of Angiotensin and Related Hormones*, ed. J. P. Buckley and C. M. Ferrario, pp. 325–56. Oxford: Pergamon.

Pickering, L. & Hogan, G. 1971. Voluntary water intoxication in a normal child. *Journal of Pediatrics*, **78**, 316–8.

Pliny the Elder (Gaius Plinius Secundus). A.D. 65–75. *Naturalis Historia*.

Porter, J. H., Young, R. & Moeschl, T. P. 1978. Effects of water and saline preloads on schedule-induced polydipsia in the rat. *Physiology and Behavior*, **21**, 333–8.

Ramsay, D. J. 1978. Beta-adrenergic thirst and its relation to the renin–angiotensin system. *Federation Proceedings*, **37**, 2689–93.

Ramsay, D. J. 1979. The brain renin angiotensin system: a re-evaluation. *Neuroscience*, **4**, 313–21.

Ramsay, D. J. & Reid, I. A. 1975. Some central mechanisms of thirst in the dog. *Journal of Physiology (London)*, **253**, 517–25.

Ramsay, D. J., Rolls, B. J. & Wood, R. J. 1975. The relationship between elevated water intake and oedema associated with congestive cardiac failure in the dog. *Journal of Physiology (London)*, **244**, 303–12.

Ramsay, D. J., Rolls, B. J. & Wood, R. J. 1977a. Thirst following water deprivation in dogs. *American Journal of Physiology*, **232**, R93–100.

Ramsay, D. J., Rolls, B. J. & Wood, R. J. 1977b. Body fluid changes which influence drinking in the water-deprived rat. *Journal of Physiology (London)*, **266**, 453–69.

Ramsay, D. J. & Thrasher, T. N. 1980. Mechanisms which terminate drinking in dogs. In *Proceedings of the Seventh International Conference on the Physiology of Food and Fluid Intake*. Warsaw.

Ramsay, D. J., Thrasher, T. N. & Keil, L. C. (1980). Stimulation and inhibition of drinking and vasopressin secretion in dogs. In *Antidiuretic Hormone*, ed. S. Yoshida, L. Share and K. Yagi, pp. 97–113. Tokyo: Japan Scientific Society Press.

Reed, M. H., Thrasher, T. N., Simpson, J. B., Keil, L. C. & Ramsay, D. J. 1980. Lesions of the subfornical organ inhibit angiotensin-induced drinking in the dog. *Society for Neuroscience Abstracts*, **6**, 531.

Reeve, E. B. & Kulhanek, L. 1967. Regulation of body water content: a preliminary analysis. In *Physical Bases of Circulatory Transport: Regulation and Exchange*, ed. E. B. Reeve and O. C. Guyton, pp. 151–77. Philadelphia: Saunders.

Richardson, D. B. & Mogenson, G. J. 1981. Water intake elicited by injections of angiotensin II into the preoptic area of rats. *American Journal of Physiology*, **240**, R70–4.

Richter, C. P. 1936. Increased salt appetite in adrenalectomized rats. *American Journal of Physiology*, **115**, 155–61.

Richter, C. P. & Eckert, J. F. 1935. Further evidence for the primacy of polyuria in diabetes insipidus. *American Journal of Physiology*, **113**, 578–81.

Rogers, P. W. & Kurtzman, N. A. 1973. Renal failure, uncontrollable thirst, and hyperreninemia. Cessation of thirst with bilateral nephrectomy. *Journal of the American Medical Association.* **225**, 1236–8.

Rolls, B. J. 1970. Drinking by rats after irritative lesions in the hypothalamus. *Physiology and Behavior*, **5**, 1385–93.

Rolls, B. J. 1971. The effect of intravenous infusion of antidiuretic hormone on water intake in the rat. *Journal of Physiology (London)*, **219**, 331–9.

Rolls, B. J. 1975. Interaction of hunger and thirst in rats with lesions of the preoptic area. *Physiology and Behavior*, **14**, 537–43.

Rolls, B. J. 1979. How variety and palatability can stimulate appetite. *Nutrition Bulletin*, **5**, 78–86.

Rolls, B. J., Jones, B. P. & Fallows, D. J. 1972. A comparison of the motivational properties of thirst induced by intracranial angiotensin and water deprivation. *Physiology and Behavior*, **9**, 777–82.

Rolls, B. J. & McFarland, D. J. 1973. Hydration releases inhibition of feeding produced by intracranial angiotensin. *Physiology and Behavior*, **11**, 881–4.

Rolls, B. J. & Ramsay, D. J. 1975. The elevation of endogenous angiotensin and thirst in the dog. In *Control Mechanisms of Drinking*, ed. G. Peters, J. T. Fitzsimons and L. Peters-Haefeli, pp. 74–8. Berlin: Springer-Verlag.

Rolls, B. J. & Rolls, E. T. 1973. Effects of lesions in the basolateral amygdala on fluid intake in the rat. *Journal of Comparative and Physiological Psychology*, **83**, 240–7.

Rolls, B. J. & Rolls, E. T. 1981. The control of drinking. *British Medical Bulletin*, **37**, 127–30.

Rolls, B. J., Rolls, E. T. & Rowe, E. A. 1981. Sensory specific satiety in man. *Physiology and Behavior*, **27**, 137–42.

Rolls, B. J., Rowe, E. A. & Rolls, E. T. 1981. The influence of variety on human food selection. In *Psychobiology of Human Food Selection*, ed. L. M. Barker, Westport, Conn: AVI Publishing Co, in press.

Rolls, B. J., Rowe, E. A., Rolls, E. T., Kingston, B., Megson, A. & Gunary, R. 1981. Variety in a meal enhances food intake in man. *Physiology and Behavior*, **26**, 215–21.

Rolls, B. J. & Wood, R. J. 1977a. The role of angiotensin in thirst. *Pharmacology Biochemistry and Behavior*, **6**, 245–50.

Rolls, B. J. & Wood, R. J. 1977b. Excretion following drinking in the dog. *Journal of Physiology* (London), **272**, 73–4P.

Rolls, B. J. & Wood, R. J. 1979. Homeostatic control of drinking: a surviving concept. *Behavioral and Brain Sciences*, **2**, 116–7.

Rolls, B. J., Wood, R. J. & Rolls, E. T. 1980. Thirst: The initiation, maintenance, and termination of drinking. In *Progress in Psychobiology and Physiological Psychology*, Vol. 9, ed. J. M. Sprague and A. N. Epstein, pp. 263–321. New York: Academic.

Rolls, B. J., Wood, R. J., Rolls, E. T., Lind, H., Lind, R. W. & Ledingham, J. G. G. 1980. Thirst following water deprivation in humans. *American Journal of Physiology*, **239**, R476–82.

Rolls, B. J., Wood, R. J. & Stevens, R. M. 1978. Effects of palatability on body fluid homeostasis. *Physiology and Behavior*, **20**, 15–19.

Rolls, E. T. 1974. The neural basis of brain-stimulation reward. *Progress in Neurobiology*, **3**, 71–160.

Rolls, E. T. 1975. *The Brain and Reward*. Oxford: Pergamon.

Rolls, E. T. 1976. Neurophysiology of feeding. *Life Sciences Research Report* (Dahlem Konferenzen, Berlin), **2**, 21–42.

Rolls, E. T. 1978. The neurophysiology of feeding. *Trends in Neurosciences*, **1**, 1–3.

Rolls, E. T. 1979. Effects of electrical stimulation of the brain on behaviour. In *Psychology Surveys*, ed. K. Connolly, Vol. 2, 151–69. London: George, Allen & Unwin.

Rolls, E. T. 1980. Activity of hypothalamic and related neurons in the alert animal. In *Handbook of the Hypothalamus*, ed. P. J. Morgane and J. Panksepp, Vol. 3A, pp. 439–66. New York: Dekker.

Rolls, E. T. 1981a. Processing beyond the inferior temporal visual cortex related to feeding, learning and striatal function. In *Brain Mechanisms of Sensation*, ed. Y. Katsuki, R. Norgren and M. Sato. New York: Wiley.

Rolls, E. T. 1981b. Central nervous mechanisms related to feeding and appetite. *British Medical Bulletin*, **37**, 131–4.

Rolls, E. T. Burton, M. J. & Mora, F. 1976. Hypothalamic neuronal responses associated with the sight of food. *Brain Research*, **111**, 53–66.

Rolls, E. T., Burton, M. J. & Mora, F. 1980. Neurophysiological analysis of brain-stimulation reward in the monkey. *Brain Research*, **194**, 337–57.

Rolls, E. T., Perrett, D. I., Thorpe, S. J., Maddison, S. & Caan, W. 1979. Activity of striatal neurons in the behaving monkey: Implications for the neural basis of schizophrenia and Parkinsonism. *Society for Neuroscience Abstracts*, **5**, 1167.

Rolls, E. T. & Rolls, B. J. 1981. Brain mechanisms involved in feeding. In *Psychobiology of Human Food Selection*, ed. L. M. Barker, Westport, Conn: AVI Publishing Co., in press.

Rolls, E. T., Sanghera, M. K. & Roper-Hall, A. 1979. The latency of activation of neurons in the lateral hypothalamus and substantia innominata during feeding in the monkey. *Brain Research*, **164**, 121–35.

Rolls, E. T., Thorpe, S. J., Maddison, S., Roper-Hall, A., Puerto, A. & Perrett, D. 1979. Activity of neurones in the neostriatum and related structures in the alert animal. In *The Neostriatum*, ed. I. Divac and R. G. E. Oberg, pp. 163–82. Oxford: Pergamon.

Rolls, E. T., Thorpe, S. J., Perrett, D. I., Boytim, M., Wilson, F. A. W. & Szabo, I. 1981. Responses of striatal neurons in the behaving monkey: influence of dopamine. *Society for Neuroscience Abstracts*, **7**, in press.

Rowland, N. 1977. Regulatory drinking: do physiological substrates have an ecological niche? *Biobehavioral Reviews*, **1**, 261–72.

Rowland, N. 1979. Natural drinking, interactions with feeding, and species differences – three data deserts. *Behavioral and Brain Sciences*, **2**, 117–8.

Rowland, N. & Nicolaïdis, S. 1976. Metering of fluid intake and determinants of ad lib drinking in rats. *American Journal of Physiology*, **231**, 1–8.

Rullier, 1821. Soif. In *Dictionaire des Sciences médicales par une Société de Médecins et de Chirurgiens*, Vol. 51, pp. 488–90. Paris: Panckoucke.

Russell, P. J. D., Abdelaal, A. E. & Mogenson, G. J. 1975. Graded levels of hemorrhage, thirst and angiotensin II in the rat. *Physiology and Behavior*, **15**, 117–9.

Sanghera, M. K., Rolls, E. T. & Roper-Hall, A. 1979. Visual responses of neurons in the dorsolateral amygdala of the alert monkey. *Experimental Neurology*, **63**, 610–26.

Schmitt, M. 1973. Influences of hepatic portal receptors on hypothalamic feeding and satiety centers. *American Journal of Physiology*, **225**, 1089–95.

Schrier, R. W. & Berl, T. 1976. Disorders of water metabolism. In *Renal and Electrolyte Disorders*, ed. R. W. Schrier, pp. 1–44. Boston: Little, Brown & Co.

Schwob, J. E. & Johnson, A. K. 1977. Angiotensin-induced dipsogenesis in domestic fowl (*Gallus gallus*). *Journal of Comparative and Physiological Psychology*, **91**, 182–8.

Segal, E., Oden, D. L. & Deadwyler, S. A. 1965. Determinants of polydipsia: IV. Free-reinforcement schedules. *Psychonomic Science*, **19**, 11–2.

Setler, P. E. 1971. Drinking induced by injection of angiotensin II into the hypothalamus of the rhesus monkey. *Journal of Physiology* (London), **217**, 59–60P.

Setler, P. E. 1973. The role of catecholamines in thirst. In *The Neuropsychology of Thirst: New Findings and Advances in Concepts*, ed. A. N. Epstein, H. R. Kissileff and E. Stellar, pp. 279–91. Washington: Winston.

Setler, P. E. 1977. The neuroanatomy and neuropharmacology of drinking. In *Handbook of Psychopharmacology*, Vol. 8, *Drugs, Neurotransmitters, and Behavior*. ed. L. L. Iversen, S. D. Iversen and S. H. Snyder, pp. 131–58. New York: Plenum.

Simpson, J. B., Epstein, A. N. & Camardo, J. S. 1977. The localization of receptors for the dipsogenic action of angiotensin II in the subfornical organ. *Journal of Comparative and Physiological Psychology*, **91**, 1220–31.

Simpson, J. B., Reid, I. A., Ramsay, D. J. & Kipen, H. 1978. Mechanism of the dipsogenic action of tetradecapeptide renin substrate. *Brain Research*, **157**, 63–72.

Simpson, J. B. & Routtenberg, A. 1972. The subfornical organ and carbachol-induced drinking, *Brain Research*, **45**, 135–52.

Simpson, J. B. & Routtenberg, A. 1973. Subfornical organ: site of drinking elicitation by angiotensin II. *Science*, **181**, 1172–5.

Smith, D. F., Balagura, S. & Lubran, M. 1970. 'Antidotal thirst': a response to intoxication. *Science*, **167**, 297–8.

Smith, G. P. 1973. Introduction: Neuropharmacology of Thirst. In *The Neuropsychology of Thirst: New Findings and Advances in Concepts*, ed. A. N. Epstein, H. R. Kissileff and E. Stellar, pp. 231–41. Washington, D.C.: Winston.

Smith, R. W. & McCann, S. M. 1962. Alterations in food and water intake after hypothalamic lesions in the rat. *American Journal of Physiology*, **203**, 366–70.

Snapir, N., Robinson, B. & Godschalk, M. 1976. The drinking response of the chicken to peripheral and central administration of angiotensin II. *Pharmacology Biochemistry and Behavior*, **5**, 5–10.

Sobocinska, J. 1969. Abolition of effect of hypovolemia on the thirst threshold after cervical vagosympathectomy in dogs. *Bulletin de l'Académie Polonaise des Sciences*, **17**, 341–6.

Sobocinska, J. 1978. Gastric distension and thirst: Relevance to the osmotic thirst threshold and metering of water intake. *Physiology and Behavior*, **20**, 497–501.

Staddon, J. E. R. 1977. Schedule-induced behavior. In *Handbook of Operant Behavior*, ed. W. K. Honig and J. E. R. Staddon, pp. 125–52. New Jersey: Prentice-Hall.

Stamoutsos, B. A., Carpenter, R. G. & Grossman, S. P. 1981. Role of angiotensin-II in the polydipsia of diabetes insipidus in the Brattleboro rat. *Physiology and Behavior*, **26**, 691–3.

Steggerda, F. R. 1941. Observations on the water intake in an adult man with dysfunctioning salivary glands. *American Journal of Physiology*, **132**, 517–21.

Stein, L. & Seifter, J 1962. Muscarinic synapses in the hypothalamus. *American Journal of Physiology*, **202**, 751–6.

Stricker, E. M. 1968. Some physiological and motivational properties of the hypovolemic stimulus for thirst. *Physiology and Behavior*, **3**, 379–85.

Stricker, E. M. 1973. Thirst, sodium appetite, and complementary physiological contributions to the regulation of intravascular fluid volume. In *The Neuropsychology of Thirst: New Findings and Advances in Concepts*, ed. A. N. Epstein, H. R. Kissileff and E. Stellar, pp. 73–98. New York: John Wiley & Sons.

Stricker, E. M. 1978. The renin–angiotensin system and thirst: some unanswered questions. *Federation Proceedings*, **37**, 2704–10.

Stricker, E. M. 1981. Thirst and sodium appetite after colloid treatment in rats. *Journal of Comparative and Physiological Psychology*, **95**, 1–25.

Stricker, E. M. & Adair, E. R. 1966. Body fluid balance, taste, and post-prandial factors in schedule-induced polydipsia. *Journal of Comparative and Physiological Psychology*, **62**, 449–54.

Stricker, E. M. & Jalowiec, J. E. 1970. Restoration of intravascular fluid volume following acute hypovolemia in rats. *American Journal of Physiology*, **218**, 191–6.

Stricker, E. M., Swerdloff, A. F. & Zigmond, M. J. 1978. Intrahypothalamic injections of kainic acid produce feeding and drinking deficits in rats. *Brain Research*, **158**, 470–3.

Stricker, E. M. & Wolf, G. 1967. The effects of hypovolemia on drinking of rats with lateral hypothalamic damage. *Proceedings of the Society for Experimental Biology and Medicine*, **124**, 816–20.

Stricker, E. M. & Zigmond, M. J. 1974. Effect on homeostasis of intraventricular injections of 6-hydroxydopamine in rats. *Journal of Comparative and Physiological Psychology*, **86**, 973–94.

Stricker, E. M. & Zigmond, M. J. 1976. Recovery of function after damage to central catecholamine-containing neurons: a neurochemical model for the lateral hypothalamic syndrome. In *Progress in Psychobiology and Physiological Psychology*, ed. J. M. Sprague and A. N. Epstein, pp. 121–88. New York: Academic.

Stuart, C. A., Neelon, F. A. & Lebovitz, H. E. 1980. Disordered control of thirst in hypothalamic–pituitary sarcoidosis. *New England Journal of Medicine*, **303**, 1078–81.

Sturgeon, R. D., Brophy, P. D. & Levitt, R. A. 1973. Drinking elicited by intracranial microinjection of angiotensin in the cat. *Pharmacology Biochemistry and Behavior*, **1**, 353–5.

Swanson, L. W., Kucharczyk, J. & Mogenson, G. J. 1978. Autoradiographic evidence for pathways from the medial preoptic area to the midbrain involved in the drinking response to angiotensin II. *Journal of Comparative Neurology*, **178**, 645–60.

Swanson, L. W. & Sharpe, L. G. 1973. Centrally induced drinking: comparison of angiotensin II- and carbachol-sensitive sites in rats. *American Journal of Physiology*, **225**, 566–73.

Takei, Y. 1977. Angiotensin and water intake in the Japanese quail (*Coturnix coturnix japonica*). *General and Comparative Endocrinology*, **31**, 364–72.

Teitelbaum, P. & Epstein, A. N. 1962. The lateral hypothalamic syndrome: recovery of feeding and drinking after lateral hypothalamic lesions. *Psychological Review*, **69**, 74–90.

Teitelbaum, P. & Stellar, E. 1954. Recovery from the failure to eat produced by hypothalamic lesions. *Science*, **120**, 894–5.

Thrasher, T. N., Brown, C. J., Keil, L. C. & Ramsay, D. J. 1980. Thirst and vasopressin release in the dog: an osmoreceptor or sodium receptor mechanism? *American Journal of Physiology*, **238**, R333–9.

Thrasher, T. N., Jones, R. G., Keil, L. C., Brown, C. J. & Ramsay, D. J. 1980. Drinking and vasopressin release during ventricular infusions of hypertonic solutions. *American Journal of Physiology*, **238**, R340–5.

Thrasher, T. N., Simpson, J. B. & Ramsay, D. J. 1980. Drinking responsiveness following ablation of the subfornical organ (SFO) or organum vasculosum of the lamina terminalis (OVLT) in dogs. In *Proceedings of the Seventh International Conference on the Physiology of Food and Fluid Intake*. Warsaw.

Toates, F. M. 1979. Homeostasis and drinking. *Behavioral and Brain Sciences*, **2**, 95–139.

Toates, F. M. 1980. *Animal Behavior – A Systems Approach*. Chichester: John Wiley & Sons.

Toates, F. M. & Oatley, K. 1970. Computer simulation of thirst and water balance. *Medical and Biological Engineering*, **8**, 71–87.

Towbin, E. J. 1949. Gastric distension as a factor in the satiation of thirst in esophagostomized dogs. *American Journal of Physiology*, **159**, 533–41.

Trippodo, N. C., McCaa, R. E. & Guyton, A. C. 1976. Effect of prolonged angiotensin II infusion on thirst. *American Journal of Physiology*, **230**, 1063–6.

Ungerstedt, U. 1971. Stereotaxic mapping of the monoamine pathways in the rat brain. *Acta Physiologica Scandinavica*, Suppl. **367**, 1–48.

Valenstein, E. S., Cox, V. C. & Kakolewski, J. W. 1970. Re-examination of the role of the hypothalamus in motivation. *Psychological Review*, **77**, 16–31.

Vander, A. J., Sherman, J. H. & Luciano, D. S. 1975. *Human Physiology – The Mechanisms of Body Function*. New York: McGraw-Hill.

Verney, E. G. 1947. The antidiuretic hormone and the factors which determine its release. *Proceedings of the Royal Society, London, Series B*, **135**, 25–106.

Vincent, J. D., Arnauld, E. & Bioulac, B. 1972. Activity of osmosensitive single cells in the hypothalamus of the behaving monkey during drinking. *Brain Research*, **44**, 371–84.

Wada, M., Kobayashi, H. & Farner, D. S. 1975. Induction of drinking in the white-crowned sparrow, *Zonotrichia leucophrys gambelii*, by intracranial injection of angiotensin II. *General and Comparative Endocrinology*, **26**, 192–7.

Walsh, L. L. & Grossman, S. P. 1973. Zona incerta lesions: disruption of regulatory water intake. *Physiology and Behavior*, **11**, 885–7.

Walsh, L. L. & Grossman, S. P. 1976. Zona incerta lesions impair osmotic but not hypovolaemic thirst. *Physiology and Behavior*, **16**, 211–5.

Walsh, L. L. & Grossman, S. P. 1977. Electrolytic lesions and knife cuts in the region of the zona incerta impair sodium appetite. *Physiology and Behavior*, **18**, 587–96.

Walsh, L. L. & Grossman, S. P. 1978. Dissociation of responses to extracellular thirst stimuli following zona incerta lesions. *Pharmacology Biochemistry and Behavior*, **8**, 409–15.

Warden, C. J. 1931. *Animal Motivation Studies: The Albino Rat*. New York: Columbia.

Weisinger, R. S. 1975. Conditioned and pseudoconditioned thirst and sodium appetite. In *Control Mechanisms of Drinking*, ed. G. Peters, J. T. Fitzsimons and L. Peters-Haefeli, pp. 148–54. Berlin: Springer-Verlag.

Weiss, C. S. & Almli, C. R. 1975. Lateral preoptic and lateral hypothalamic units:

in search of the osmoreceptors for thirst. *Physiology and Behavior*, **15**, 713–22.

Wettendorf, H. 1901. Modifications du sang sous l'influence de la privation d'eau: contribution à l'étude de la soif. *Travaux du Laboratoire de Physiologie, Instituts Solvay*, **4**, 353–484.

Williams, D. R. & Teitelbaum, P. 1956. Control of drinking behavior by means of an operant conditioning technique. *Science*, **124**, 1294–6.

Winson, J. & Miller, N. E. 1970. Comparison of drinking elicited by eserine or DFP injected into preoptic area of rat brain. *Journal of Comparative and Physiological Psychology*, **73**, 233–7.

Wirth, J. B. & Epstein, A. N. 1976. Ontogeny of thirst in the infant rat. *American Journal of Physiology*, **230**, 188–98.

Wolf, A. V. 1950. Osmometric analysis of thirst in man and dog. *American Journal of Physiology*, **161**, 75–86.

Wolf, A. V. 1958. *Thirst; Physiology of the Urge to Drink and Problems of Water Lack*. Springfield, Ill: Thomas.

Wood, R. J., Maddison, S., Rolls, E. T., Rolls, B. J. & Gibbs, J. 1980. Drinking in the rhesus monkey: roles of pre-systemic and systemic factors in drinking control. *Journal of Comparative and Physiological Psychology*, **94**, 1135–48.

Wood, R. J., Rolls, B. J. & Ramsay, D. J. 1977. Drinking following intracarotid infusions of hypertonic solutions in dogs. *American Journal of Physiology*, **232**, R88–92.

Young, C. E. & McDonald, I. R. 1978. The effect of intravenous infusion of angiotensin II on drinking in the Australian marsupial *Trichosurusvulpecula*. *Journal of Physiology* (London), **280**, 77–85.

Zigmond, M. J. & Stricker, E. M. 1973. Recovery of feeding and drinking by rats after intraventricular 6-hydroxydopamine or lateral hypothalamic lesions. *Science*, **182**, 717–20.

Zimmerman, M. B., Blaine, E. H. & Stricker, E. M. 1981. Water intake in hypovolemic sheep: effects of crushing the left atrial appendage. *Science*, **211**, 489–91.

Index